MW00616761

APPLIED FORMAL VERIFICATION

APPLIED FORMAL VERIFICATION

DOUGLAS L. PERRY
Virtutech, Inc.

HARRY D. FOSTER
Jasper Design Automation, Inc.

McGraw-Hill

New York Chicago San Francisco Lisbon London Madrid
Mexico City Milan New Delhi San Juan Seoul
Singapore Sydney Toronto

The *McGraw-Hill* Companies

Cataloging-in-Publication Data is on file with the Library of Congress

1 2 3 4 5 6 7 8 9 0   DOC/DOC   0 1 0 9 8 7 6 5

ISBN 0-07-144372-X

The sponsoring editor for this book was Stephen S. Chapman, the editing supervisor was Stephen M. Smith, and the production supervisor was Pamela A. Pelton. It was set in Palatino by Wayne A. Palmer of McGraw-Hill Professional's Hightstown, N.J., composition unit. The art director for the cover was Anthony Landi.

Printed and bound by RR Donnelley.

McGraw-Hill books are available at special quantity discounts to use as premiums and sales promotions, or for use in corporate training programs. For more information, please write to the Director of Special Sales, McGraw-Hill Professional, Two Penn Plaza, New York, NY 10121-2298. Or contact your local bookstore.

 This book is printed on recycled, acid-free paper containing a minimum of 50% recycled, de-inked fiber.

C O N T E N T S

Chapter 7

The Formal Test Plan Process   121

Chapter 8

Techniques for Proving Properties   135

PREFACE

There are many excellent technical books on formal verification technology and algorithms. However, the field lacks a book focused on applied formal verification. This book was written to help hardware design engineers learn how to apply formal verification to real-world design problems. The goal of this book is to amass enough formal verification information to allow design engineers to quickly verify their designs using its techniques. No former knowledge of formal verification is necessary to understand the concepts we present; however, knowledge of VHDL or Verilog and simulation-based verification will be useful. You will not learn everything there is to know about formal verification, but enough to be able to efficiently apply formal verification techniques to a real design.

This book is divided into two logical sections. The first section introduces general verification techniques in use today and compares them with formal verification techniques. Through this process, you will learn how formal verification differs from traditional simulation approaches to design verification. The second section gives details about formal verification techniques. The focus is on creating formal high-level requirements: how to write them and apply them to your design.

In the first section, Chaps. 1 to 3, we present the reasons behind verification and the typical process for verification. This will give the reader a foundation for the rest of the book. Chapter 1 introduces the importance of verification. Chapter 2 discusses the typical techniques used in simulation-based verification, the most prevalent method of functional verification at the time of this writing. Chapter 3 describes the most common techniques for performing verification today and the tradeoffs that each technique requires. This chapter will be helpful for determining when to use each type of tool in a typical verification flow.

The second section, Chaps. 4 to 9, introduces applied formal verification. Chapter 4 discusses formal verification concepts for both applied boolean and sequential verification. In Chap. 5 we present a set of definitions commonly used in the field. Chapter 6 introduces formal property checking and discusses the emerging PSL and SystemVerilog assertion standards. Building on this knowledge, Chap. 7 discusses the process of creating a formal test plan. Chapter 8 introduces various advanced techniques that are useful when a formal tool encounters state explosion. Finally, Chap. 9 discusses formal verification in the context of the full system verification process.

There are two appendices with tabulated reference information. Appendix A contains a table of commonly used PSL statements for high-level requirements. Appendix B contains a table of similar high-level requirements specified in SystemVerilog syntax.

Douglas L. Perry
Harry D. Foster

APPLIED
FORMAL
VERIFICATION

Introduction to Verification

The number of transistors that can be contained on a silicon device continues to increase year after year. While many think that the silicon limits are approaching in the near future, today's devices continue to be extremely complex to design and create error-free.

Building a device that contains hundreds of millions of transistors is a nontrivial task that requires a deliberate and consistent verification methodology to prevent design errors. This chapter will focus on some of the reasons that design verification is so important.

A design containing hundreds of millions of transistors will typically be split into smaller functional units or blocks, to manage the complexity. The entire design will be composed of hierarchical levels of blocks, as shown in Fig. 1.1.

The device is a single-chip CPU device with the CPU core, on-chip RAM, on-chip ROM, Ethernet MAC, and HDLC controller. Each block contains a smaller autonomous piece of the entire design functionality. These blocks are interconnected via signals and busses to form the complete design.

When designs were much smaller, one designer had the ability to manage the entire design. However, with the complexity available

FIGURE 1.1

Simple Chip Design

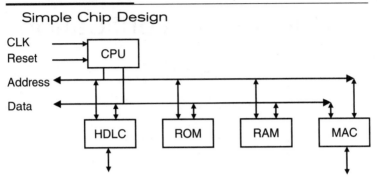

now, this would be extremely difficult or impossible. Therefore each designer manages the creation of one or more blocks of the design.

To enable proper operation of the entire design, the block interconnections must be correct and the blocks themselves must be correct. Depending on the design team, the designers may have access to a written microarchitectural specification within which the functionality of a particular designer's block is described in detail. Detailed block interconnect specifications complete the functional description of the design and allow the designer of a particular block to determine the interface requirements of her or his block with other blocks.

1.1 VERIFICATION

Designers create blocks that match the block behavior in the microarchitectural specification and the block interconnections specified by the block interface specification. However, subtleties always exist in complex designs that cause unexpected behavior when the design is actually built. How does a designer determine that the design RTL description behaves as expected? This is an increasingly important question as design complexity increases. There is a nonlinear relationship between design complexity and verification complexity, as shown in Fig. 1.2. As design complexity rises, the verification complexity increases faster, making it even harder to verify a design as the design size increases.

FIGURE 1.2

Classic Logic System

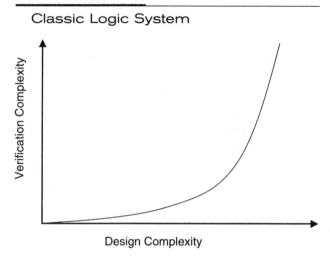

Design Complexity (x-axis), *Verification Complexity* (y-axis)

The engineering costs associated with creating a device using 90 nm (nanometers), 65 nm, or below are staggering. Design errors not only create costs in reengineering but can also drastically impact the time to market. Now more than ever, getting a design correct the first time is very important not only to save money on a project but also to keep a company financially sound.

As device geometries shrink, it is important that the designer verify the design before it is built. Verification is the process of testing to see if the device behaves properly in the target environment. Ideally the designer would build the real device, place it in the target environment, and see if it works. However, there are a number of reasons why this method does not work well for ASIC design.

First, the cost of building the actual device is such that a typical project can afford only one or two versions of the device. To build a new device, new masks are required. Each new mask set can cost hundreds of thousands of dollars. The device has to be nearly perfect the first time. As already discussed, each version of the device has high engineering and time-to-market costs.

Second, there is a lack of observability into the device. A typical ASIC device has hundreds of thousands or millions of signals, yet only a few hundreds or thousands of pins. This means that only a tiny percentage of the signals of the design are visible during debugging.

A number of techniques are used to try to get greater visibility. These include dedicated debug busses that multiplex internal signals to external pins. While this does provide greater visibility into the device, the number of signals that a designer is able to see at once is limited by the number of external pins. Also the additional multiplexer logic can cause excess delay in the target design or can introduce errors.

Finally, to build prototypes of devices and to test those devices to determine design correctness will be very expensive per device, and they will be difficult to debug. The lowest-cost solution is always to verify that the design is correct before building it and to build it only once. This also prevents a design from missing its market window.

1.2 MARKET WINDOW

Electronic devices typically have a fixed market window into which the device can be sold and the company can make a reasonable profit on the engineering investment. It is very important that the device be available during the market window for that device, or else sales will be poor. A market window for a device depends on a number of factors that are not covered in this book.

Sales of a device are usually represented by a chart, as shown in Fig. 1.3. The horizontal scale is time, the vertical scale is the instantaneous revenue. Time starts when the first production units start shipping. Until the product gains wide acceptance, the revenue rises slowly. Once the product becomes widely accepted, the curve rises rapidly. At some point the product sales will peak. Market saturation, competition, or other factors will start to erode the market, and sales will begin to fall.

Figure 1.4 shows a graph of cumulative product revenue over time. The horizontal axis shows time, and the vertical axis shows total accumulated revenue. The revenue starts accumulating slowly until the product becomes accepted in the market. Once it is accepted, the total revenue rises rapidly. This increase continues until sales start to drop off. Once sales start to drop off, the total revenue increases much more slowly.

FIGURE 1.3

Product Revenue over Time

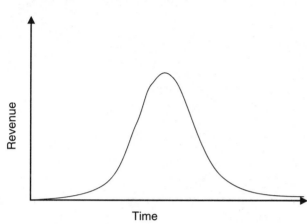

FIGURE 1.4

Cumulative Product Revenue

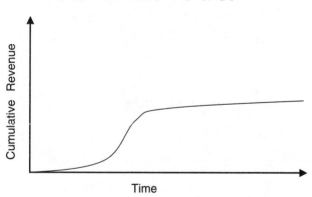

Let's take a look at what happens when the product arrives to market late. In Fig. 1.5 the dotted line represents the product arriving to the market later. Sales start off more slowly, the sales peak occurs later, and the peak is much less steep than if the product were delivered on time.

The effect on total revenue is shown in Fig. 1.6. The dotted line represents the total revenue of the product arriving late to market. As can be seen, the accumulated total revenue rises later and never

FIGURE 1.5

Product Late to Market

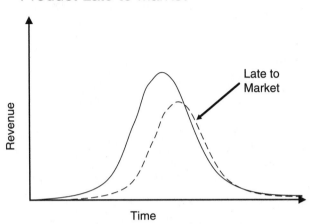

FIGURE 1.6

Product Late to Market

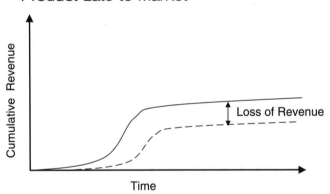

reaches the same level as that of the product delivered on time. The difference between the two lines represents the revenue shortfall caused by not getting the product to market on time and missing the market window.

Depending on development costs it's possible that the company may make very little or no profit on a late product; in fact the company may actually lose a significant amount of money. Therefore a designer must minimize all the risks in getting a product to market on time.

1.3 SUMMARY

The largest risk is that the product does not work when delivered. A rigorous verification methodology ensures that the product does work, and can help prevent delays that cause a product to miss the market window.

CHAPTER 2

Verification Process

Chapter 1 described why it is so important that the product be delivered on time. To get a product to market faster, the design team must be convinced as soon as possible that the design will work. The design team will focus a tremendous amount of effort to prove to themselves that the design is correct by verifying the design. This chapter will focus on the process used to verify that a design is correct.

2.1 VERIFICATION PLAN

The first step in the verification process is to develop a plan to verify the design. This is commonly referred to as a *verification plan*. This plan will identify the areas of the design to verify and how they will be tested. This is usually a free-form text document. Later chapters will explore more rigorous techniques for developing verification plans.

A verification engineer will usually work with the design engineer and the design specifications to determine the types of tests to create. These tests will be documented in the verification plan along with which tests should be executed first, second, etc.

The description of each test will usually contain the verification goals and how to run the test.

2.2 DEBUG CYCLE

The verification engineer or design engineer will run a set of tests to verify that the design is functionally correct. Tests can vary widely based on the functionality being tested. For instance, if the device is a networking chip, tests that try different kinds and types of packets will be executed. If the device is a processor, tests that try different kinds of instructions, operating systems, and application programs will be run.

As each test is developed, it is run against the target device, as shown in Fig. 2.1. The test is created and configured for the device

FIGURE 2.1

Debug Cycle

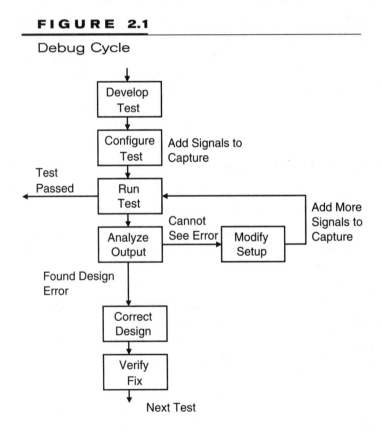

Next Test

to be tested. The test is run and the results are analyzed. If the results reveal an error, the designer will need to try to find the cause of the error. Typically the designer will not have examined the correct signals to show the cause of the error. The designer will modify the setup of the test to add more signals that the designer thinks might show the cause of the error. The test is run again and the results are analyzed. If the results show the cause of the error, the designer will fix the design or the testbench and run the test again to verify that the fix removed the error. If the test passed without any error, the verification engineer will move on to the next test.

For small tests the time to run a test is usually small. For larger tests runtimes can be enormous, taking hours or even days. Verification can quickly become the dominant factor in getting a new device to market. When lots of tests are required to verify a design, the total time of these tests can quickly dominate the time required to create the design.

2.3 SIMULATION

Simulation is one of the most common methods of verification used since the 1960s. Early simulators simulated the design at the transistor, then the gate level. Most of the simulation at the time of this writing is done at the *register transfer level* (RTL), in which the designer describes the registers in the design and the logic between the registers. Most of the designs are created using a *hardware description language* (HDL) such as VHDL or Verilog. Simulators are software applications that run on host computers such as Windows desktops and laptops, UNIX workstations, and Linux servers.

A typical RTL simulation environment looks as shown in Fig. 2.2. The simulator reads the device model and converts it to internal data structures representing the design behavior. These descriptions create a virtual design representation in the simulator. The device model typically does not contain design stimulus. Stimulus is usually applied externally to the model. This is advantageous because it allows the designer to change design stimulus without changing the model. Different design stimulus will drive the model differently and test different design functionality.

FIGURE 2.2

Typical Simulation Environment

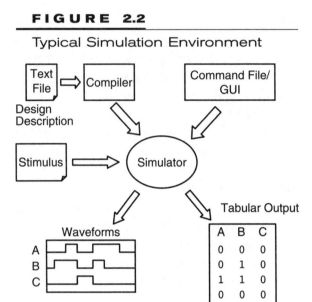

Stimulus can take a number of forms including data read from a file, a model that represents the target system environment, and combinations of the above. Stimulus is usually created by the designer or by a separate verification team. Typically the designer will create stimulus to allow the device to be tested at a basic level. The design is then passed to a verification team for rigorous verification.

2.4 OUTPUT DATA

The simulator will drive the stimulus to the design, and the simulator will model the effects of the stimulus to the design and the reaction of the design to the stimulus. Depending on user setup the simulator will capture output data from the simulation of the design. These output data can take a number of forms depending on user setup. The most common forms of data captured are tabular data representing signal values over time. These data can be captured when a signal changes value or sampled at a consistent interval.

The other form of data captured from the design is waveform data. Waveforms are simply a graphical method to display signal

FIGURE 2.3

Sample Waveform

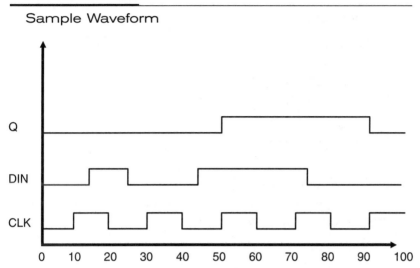

values over time. Sample waveforms of the inputs and outputs of a D flip-flop are shown in Fig. 2.3. The flip-flop is rising-edge-triggered. At time 50 the flip-flop will capture the 1 value on DIN and reflect that value on output Q. This value will be maintained until time 90 where a 0 value on DIN is captured.

When the number of signals is small and the complexity of the design is minimal, the designer can easily verify the design by visual inspection. However, as the design complexity increases, this is no longer possible, especially as the designer runs multiple tests over and over while fixing design errors. Therefore most designers will create a testbench to drive the design and monitor the results of the design. The results of the design are compared with expected results. A mismatch between the expected results and the design output means that the test failed. Incorrect stimulus was presented to the design, the design is incorrect, or the expected results are incorrect. The designer will need to verify which one(s) is (are) causing the test to fail.

A testbench is represented graphically as shown in Fig. 2.4. The testbench effectively wraps around the *device under test* (DUT). The testbench drives the DUT with stimulus and captures the results of the simulation. The results are compared with the expected results

FIGURE 2.4

Testbench Structure

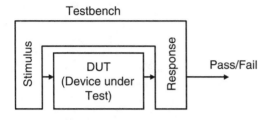

in the testbench, and a pass or fail indication is generated. If the output of the DUT matches the expected results, a pass indication is generated; otherwise, a fail is returned.

2.5 TESTBENCH DEVELOPMENT

Testbench development is a nontrivial task for complex devices. The designer must develop stimulus and expected results for all possible modes of operation for a particular design. As design complexity increases, the number of possible input scenarios increases dramatically.

A sample design process is shown in Fig. 2.5. In this process the designer will create a design specification from which the HDL design code is written. After the HDL code has been written, the designer will create some simple tests to verify that the design behavior matches the expected functionality.

If the company has a verification team, the design HDL code and specifications are given to the verification team to perform the verification. After verification the HDL code is synthesized to the target ASIC technology. Finally the place and route software is

FIGURE 2.5

Sample Design Process

Design Spec	Write HDL	Verify	Synthesis	Place Route

used to create the layout database that is used to create the masks for the actual ASIC fabrication.

As you can see, the verification step takes the most time to complete by far. Reducing the verification time can significantly decrease the time to market.

2.6 SUMMARY

Millions of dollars of research have been invested in methods to decrease the verification time, yet deliver a functionally correct design on the first try. Chapter 3 will contrast the different verification techniques used to verify a design.

Current Verification Techniques

This chapter will contrast and compare the different verification methods in use today. The background of each technique is described, and then the strengths and weaknesses of each are compared.

3.1 HDL SOFTWARE SIMULATORS

An HDL software simulator is a software application program that runs on a typical engineering workstation or PC. The HDL software simulator reads a hardware description language input file that describes the functional operation of the design. As the simulator executes, it applies stimulus and user commands to the design and generates output data for analysis. This is shown in Fig. 3.1.

An HDL description usually describes design behavior in terms of processes that run when triggered by input signal changes, as shown in Fig. 3.2.

A list of processes that need to be run is generated each time cycle; then each process is run one at a time, one after the other. Figure 3.3 shows how the HDL software simulator will execute processes. If an event occurs that triggers process 1, that process will execute and cause process 2 to execute during the next cycle. Process 2 will trigger process 3 and process 7 to execute the next cycle.

FIGURE 3.1

HDL Software Simulator

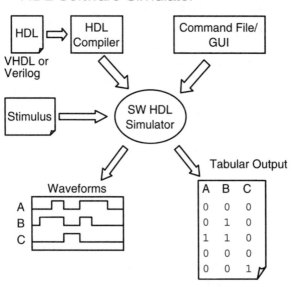

FIGURE 3.2

Process Changes

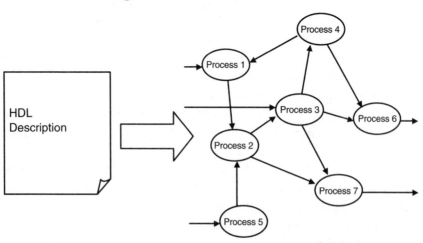

FIGURE 3.3

Process Changes

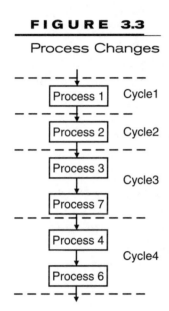

This is a very natural way to view the operation of the design, except that in the real hardware all these operations happen concurrently. Each process is also implemented by transistors in the real hardware, which also are operating concurrently. The simulator provides a functionally accurate view of the design, but the real hardware runs orders of magnitudes faster because all operations are being executed in parallel.

The Good News

HDL software simulators are relatively low-cost and can handle very large designs. Their biggest drawback is slow runtime. HDL software simulators typically cannot be interconnected to real hardware environments because the simulation speed is too slow. The simulated model is effectively data that are being executed by the simulator on the host computer, so there are no "device pins" to connect to the external environment.

For small designs, designers can think of and generate test cases for the most important scenarios to test. As the design increases in size, this process becomes increasingly difficult. An HDL software simulator will typically run a few tens of clocks per second on a fast engineering workstation. For tests that take hundreds of millions or billions of cycles to complete, the simulation runtime can take days or weeks. This is not acceptable for most chip design cycles as it adds a lot of time to the design cycle and limits how fast the design can be completed. Therefore a tremendous amount of effort has been put into increasing the speed of simulation.

Advantages of HDL Software Simulation

When one compares HDL software simulation to the real hardware implementation, a number of advantages appear. As already discussed, the real hardware implementation can be very expensive to manufacture. A simulation model takes time to create, but in the end is just a textual description usually created by a text editor.

The simulation model provides full visibility. Any signal value in the design can be examined at any time, whereas in the real hardware system only the external pins of the chip can be examined. With the real hardware system it can be very difficult to capture the pin values at the correct time. With a simulation model any signal or pin can be examined at any time during the simulation.

The simulation model allows the designer full control over the signal values in the design. At any time the designer can change the value of an internal signal. In the real design it is extremely difficult or impossible to modify the internal states of a design.

The execution of the simulation model can be stopped at any time. An RTL simulator lets the user run forward in time as little or as much as the designer requests. In the simulator the designer can run forward a small fraction of a clock cycle, a complete clock cycle, or millions of clock cycles. In contrast, the real hardware is always running, and control over starting or stopping the design is usually impossible.

The simulated model can save all the states of the simulation in a checkpoint file. The checkpoint file can be loaded back into the simulator to restore the simulator to the exact state when a checkpoint was taken. This allows the designer to restore the simulation to any time that a checkpoint was taken. This makes debugging using a simulator much easier. The designer can run the simulation to a point before an error occurs, then run forward and capture signal values from different points within the design to narrow down and find the cause of the error. The design can keep returning to any checkpoints and running forward from those points. The designer can capture signal values from different areas of the design, change signal values, and run forward to see the effects of the signal value changes.

The real hardware system can be restarted only from the reset state. Of course the real system runs orders of magnitudes faster than the RTL simulator, so arriving at a particular state may be very fast, but the real device still lacks the internal visibility needed for effective debugging.

The simulated system can also have impossible scenarios presented. The designer can insert state conditions that would be very difficult or take an extraordinary amount of time in the real hardware system. The designer can verify that the simulated system will respond correctly even in extreme conditions.

Table 3.1 compares implementing the device in real hardware, such as an ASIC technology, to using an HDL software simulator to verify the design first. As can be seen, the real hardware has the

TABLE 3.1

Real Hardware versus HDL
Software Simulator

	Real Hardware	HDL Software Simulator
Speed	Real time	Up to 10 clocks/s
Visibility	Poor	Excellent
Compile time	Poor	Fast
Debugging tools	Poor	Excellent
Checkpoint	No	Yes
Cost	Very high	Low
Testbench required	No	Yes
Simulation coverage	High	Low

highest speed and the highest coverage, but unless the chip has minimal functionality, the chances of getting it right without verification are extremely small. The HDL software simulator does have poor performance, but it does allow the user to verify important areas of the design a piece at a time in a very user-friendly manner.

3.2 ACCELERATED SIMULATION

Accelerated simulation attempts to solve the problem of slow HDL software simulator simulation speed. As discussed earlier, HDL simulation can be very slow, allowing a few tens of cycles per second on a typical large design. When millions or billions of cycles are needed to execute a particular scenario, using HDL software simulation can result in extremely long runtimes.

3.2.1 Increasing Simulation Speed

Millions of dollars of research have been put into techniques for increasing the speed of simulation. Two different types of approaches have been used to increase simulation speed: hardware approaches and software approaches.

Software Approaches

Software approaches to increase simulation speed usually involve optimizations in which some aspect of the simulation accuracy is ignored. For instance, cycle-based simulation ignores any gate delay information present in the design. Only state changes on clock cycle boundaries are taken into account. These approaches work if the accuracy factor is not important in the simulation. While these approaches will speed the simulation by 5 to 10 times for some designs, this is still not enough speed to significantly shorten the debug cycle when billions of cycles need to be run.

Hardware Approaches

Applications that require even greater simulation speed resort to hardware approaches. These approaches use some custom-built hardware to increase simulation speed. A typical simulation is a set of interconnected processes that trigger one another based on external or internal events. Internal events are usually caused by state machines or clocks. External events are caused by testbenches or other models connected to the design.

Events cause processes to be executed to recalculate their output states. In an RTL simulator each process is executed serially one after the other. This happens because the computer on which the HDL software simulator is executed has only one processor, and runs one process at a time. Hardware accelerators attempt to run as many processes as possible in parallel to increase simulation speed. Since hardware accelerators have lots of processors, lots of processes can be scheduled in parallel to increase the overall throughput.

Looking at the example used earlier on a hardware accelerator, we see that processes 3 and 7, and 4 and 6, can be executed in parallel. This is shown in Fig. 3.4. A real design would have many more processes and many more processors, so the parallelism would show much more dramatic improvements in speed.

FIGURE 3.4

Parallel Processing in a Hardware Accelerator

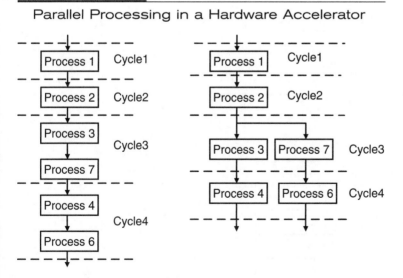

Attempts have been made to run RTL simulation on multi-processor workstations to increase simulation speed. Simulation speed is increased but not linearly with the number of processors. The RTL simulation process granularity typically has not matched the CPU granularity, and the communication between processes has become the bottleneck. Therefore the speed of simulation has been limited by interprocess communication or memory bandwidth of shared memory. Partitioning all the processes across the CPUs for high performance is also a very difficult task. Typical speedups based on the number of CPUs are very nonlinear and are similar to those shown in Fig. 3.5.

Further attempts at increasing simulation speed have involved building small custom processors with FPGA devices and using other FPGA devices to interconnect the FPGA processors. Each FPGA processor executes a small portion of the total simulation. All the processors execute in parallel, thereby increasing the speed of simulation.

FIGURE 3.5

Multi-CPU Speedup

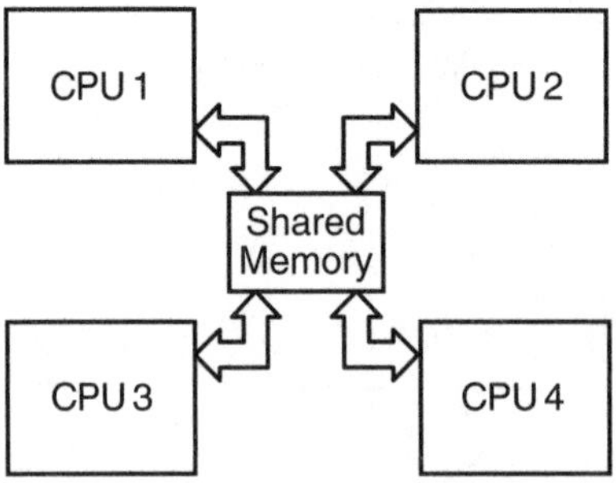

CPUs	Speedup
1	1
2	1.3
4	2.2
8	2.5

FIGURE 3.6

FPGA Hardware Accelerator

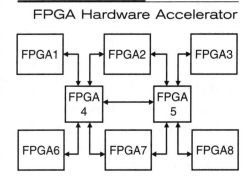

Figure 3.6 shows a block diagram of an FPGA hardware accelerator and the connections between processors. Each of the six processors executes a small piece of the total simulation. FPGAs 4 and 5 make the appropriate connections to route signals between the processors.

The Good News

These accelerators run much faster than HDL software simulators. Speeds of 75,000 to 100,000 clocks per second are possible. Accelerators typically retain all the debugging capabilities of the HDL software simulator, including output of waveforms, and trace databases. Compile times are fast and usually predictable with the size of the design.

The Bad News

The main bad-news item of accelerators is that they are usually much more expensive than an HDL software simulator. Some of these accelerators can cost 10 to 20 times the cost of an HDL software simulator. For design groups that need the extra simulation speed, the cost is justified.

One of the reasons that these accelerators are not as fast as they could be is that the underlying hardware technology is FPGA-based. FPGA-based processors will run at FPGA speeds, which are much slower than those of an ASIC or full custom design. The interconnection between FPGA processors also uses FPGA devices. This causes extra delay in the connections between FPGA processors. Both of these issues cause the accelerator to be much slower than it could potentially be.

3.2.2 Testbench

The accelerator still suffers from the same testbench issue as the HDL software simulator. The hardware accelerator does not run fast enough to connect to the real hardware environment in which the final product will exist. Therefore the hardware accelerator will need to be driven by a testbench. Most accelerators only accept synthesizable HDL. Most testbenches are written using nonsynthesizeable HDL because it makes writing the testbenches much easier. Therefore the hardware accelerator is connected to a workstation running the testbench in an HDL software simulator, as shown in Fig. 3.7. The two simulators communicate to perform the simulation. However, depending on the speed differences between the HDL software simulator and the hardware accelerator, the accelerator will typically be waiting for the testbench to execute, and therefore the entire process will run only as fast as the HDL software simulator.

Table 3.2 summarizes the differences between a typical HDL software simulator and a hardware accelerator. The main benefit of a hardware accelerator is much higher simulation speed, up to 100,000 clocks per second. The main disadvantage of a hardware accelerator is the high cost of purchase and ownership.

FIGURE 3.7

Workstation Connected to Accelerator

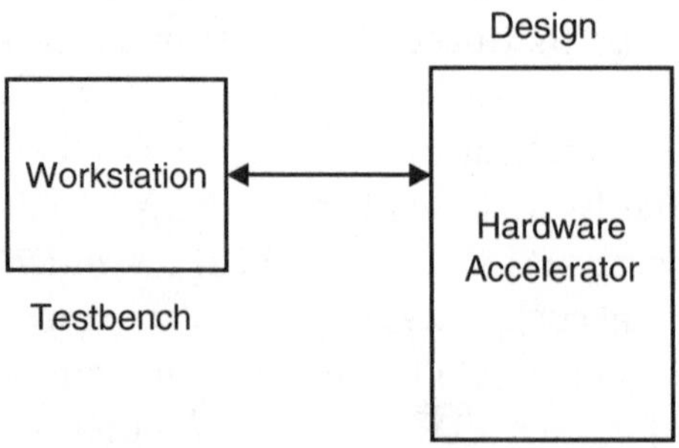

TABLE 3.2

Comparison with Hardware Accelerator

	Real Hardware	HDL Software Simulator	Hardware Accelerator
Speed	Real time	Up to 10 clocks/s	Up to 100,000 clocks/s
Visibility	Poor	Excellent	Excellent
Compile time	Poor	Fast	Fast
Debugging tools	Poor	Excellent	Excellent
Checkpoint	No	Yes	Yes
Cost	Very high	Low	Medium to high
Testbench required	No	Yes	Yes
Simulation coverage	High	Low	Low to medium

3.3 PROCESS-BASED ACCELERATOR TECHNIQUES

To further increase the speed of accelerators, some companies have used fast ASIC processors to execute the simulation. The processors are highly optimized and run at very high speed, and there are hundreds or thousands of them. These processors are connected in a grid or matrix, as shown in Fig. 3.8, and the simulation processes are distributed among the processors. The simulation processes communicate between themselves through the interface busses between the processors.

The Good News

Processor-based accelerators usually have much higher-speed processors than FPGA-based systems because they use ASICs. While FPGA-based processors might run at 100 to 150 MHz, processor-based accelerators might run at 800 MHz to 1 GHz or more. Compile times are relatively fast even for very large designs. Processor-based

FIGURE 3.8

ASIC Processor-Based
Accelerator

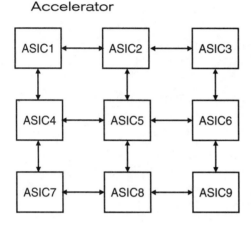

accelerators execute much faster than other hardware accelerators, typically reaching speeds of 100,000 to 500,000 clocks per second. This speed allows tests that took hours to run on an HDL software simulator to run in seconds.

The Bad News

All this speed comes at a tremendous cost. Processor-based accelerators are very expensive because of all the hardware necessary to build them. These machines typically consist of a number of very expensive boards containing lots of ASIC processor devices in a large cabinet.

While these accelerators run very fast, they still run orders of magnitudes slower than real hardware systems. Even tests that use a small amount of "real-time" data can take hours or days to run. Some designers have slowed their real hardware environment to allow connection to the real hardware. This lets the designer use "real stimulus" for some tests. The designer will still typically need to run with a testbench to cover corner cases.

Table 3.3 summarizes the differences in the verification methods discussed to date. The main differences between a hardware accel-

TABLE 3.3

ASIC Hardware Comparison

	Real Hardware	HDL Software Simulator	Hardware Accelerator	ASIC Hardware Accelerator
Speed	Real time	Up to 10 clocks/s	Up to 100,000 clocks/s	Up to 500,000 clocks/s
Visibility	Poor	Excellent	Excellent	Excellent
Compile time	Poor	Fast	Fast	Fast
Debugging tools	Poor	Excellent	Excellent	Excellent
Checkpoint	No	Yes	Yes	Yes
Cost	Very high	Low	Medium to high	High
Testbench required	No	Yes	Yes	Probably
Simulation coverage	High	Low	Low to medium	Low to medium

erator and an ASIC hardware accelerator are cost and simulation speed. Simulation speed is much higher, but so is the cost, reaching millions of dollars for large systems. Most designers will use a testbench with the ASIC hardware accelerator, but there are cases where connecting to the real hardware environment is possible.

3.4 HARDWARE EMULATION

Hardware emulators were developed in the mid-1980s when FPGA devices finally were large enough to accommodate enough logic to make emulators feasible. FPGA devices are reprogrammable logic devices that configure their functionality based on a downloaded data stream.

Usually on power-up an FPGA device will download the data stream into on-chip static RAM, as shown in Fig. 3.9. The data in the RAM are used to configure the operation of lookup tables and the

FIGURE 3.9

Data Loaded into an FPGA

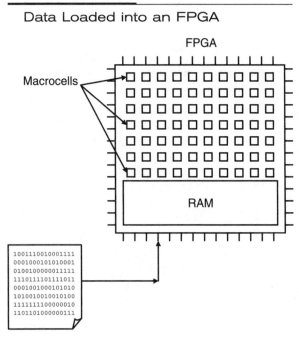

signal routing on the FPGA. Lookup tables or macro cells perform boolean operations on input signals to the tables. These boolean operations are the functional behavior of the design as specified by the designer. Another function of the RAM data is to configure routing switches to interconnect the functional lookup tables. FPGA devices contain routing channels between lookup tables with switches to connect outputs of lookup tables to these routing channels. When a switch is turned on, the output signal of the lookup table is connected to a wire in the routing channel. Another switch will connect the routing channel signal to the input of another lookup table. The complete functionality of the design is implemented by configuring the lookup tables with the correct boolean function and interconnecting the lookup tables with the correct topology.

The main drawbacks to this technology are speed and price. While the lookup tables themselves are quite fast, the signal connections between lookup tables typically limit the speed of the design. Each signal being routed from one block to another must traverse numerous switches before the connection is made to the target lookup table. In contrast, in an ASIC device, these connections are simply wires from one logic gate to another.

The FPGA device requires much more logic to implement the same functionality of an ASIC device. The FPGA device has programmable lookup tables instead of simple logic gates. The FPGA device has routing switches instead of simple wires. The FPGA device has a large configuration RAM, and all the routing from the RAM to the lookup tables and routing switches, while the ASIC has none of this. All this extra logic creates a very large silicon area penalty. Therefore the cost of an FPGA implementation from a silicon area point of view is much higher than that of an ASIC device. However, the ultimate flexibility and ease of creation still make it a very attractive option to thousands of designers. Only those devices for which cost and/or speed is the driving factor are not good candidates for an FPGA implementation.

An emulator builds on the flexibility of the FPGA device and takes the configurability of the FPGA one step further from a single chip to an entire system. The configuration RAM of the FPGA device configures the FPGA device only. The emulator uses a number of circuit boards containing multiple FPGA devices, where all the device configurations are used to implement the entire system.

Some of the FPGA devices are used for the logic and memory of the design. Other FPGA devices are used for signal routing between. The emulator uses special software to synthesize the user HDL design to logic gates. These logic gates are partitioned into multiple FPGA devices. The routing between devices is calculated, and the configuration RAM of the routing FPGA devices is loaded with the necessary configuration information to route the signals.

This is a very difficult process as the number of signals that connects one FPGA to another can be very large. The number of signals that need to connect from one FPGA to another can easily become more than the available pins on the FPGA devices. The software may try different partitions to see if better cuts can be made, or another approach called *pin multiplexing* is used, as shown in Fig. 3.10.

FIGURE 3.10

Pin Multiplexing

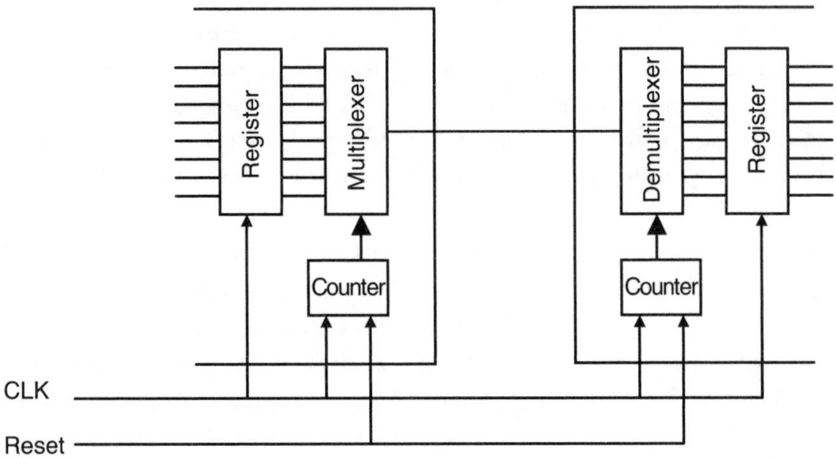

Each pin connecting one FPGA to another is time-division-multiplexed to effectively create more virtual pins. For instance, as shown in Fig. 3.10, eight pins connect from the internal FPGA logic to a register and then to an 8-to-1 multiplexer. A single signal connects from the output of the multiplexer to the pin of the FPGA device. That pin connects to the input of a demultiplexer on the second FPGA, whose outputs are registered with another register. Two counters control the states of each multiplexer and demultiplexer. The counters advance each clock cycle to give the next time slot access to the FPGA pin to transfer data.

At each clock cycle the counter will select a different input of the multiplexer and transfer it to the physical pin. At the receiving FPGA, each clock cycle will cause the counter to transfer the incoming signal to one of the eight output pins.

As seen by the timing diagram in Fig. 3.11, each pin has a finite time interval in which to transfer signal data. The effect is that a single pin acts as eight separate pins. However, if the original transfer speed of the pin was 80 MHz, each pin can now only run at 10 MHz. This is so because each pin will get one-eighth of the bandwidth of the FPGA pin.

FIGURE 3.11

Pin Multiplexing Timing

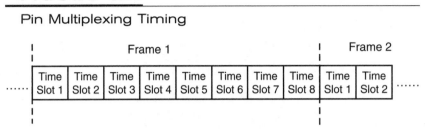

The Good News

An emulator runs fast because all the gate logic in the design is executed in parallel instead of one at a time, as in a software RTL simulator. If the design can be partitioned such that it fits into the emulator properly, good runtimes can be achieved. In fact, engineers have connected emulators into the real environment by matching the speed of the emulator in the real environment. This can work for some designs but not for others. When it does work, it allows the designer to see how his or her design will behave in the real world.

Emulators have full visibility into the design so that the designer can examine any signal at any time. This is very important during debugging.

The Bad News

Emulators are complex pieces of machinery requiring huge investments in hardware. They are extremely expensive as they use many of the largest and most expensive FPGA devices per board, and many boards per system.

Getting a design into an emulator can also be extremely challenging. Partitioning a design so that it fits into the emulator properly is very hard. The compile times of large designs are 24 h or more. This means that a typical debug cycle iteration becomes much more than a day. This can greatly extend the design cycle as usually hundreds of bugs are found in a design while implementation is in process.

Table 3.4 summarizes the different verification techniques discussed so far. The emulator provides high speed of simulation, but

TABLE 3.4

Comparison with Emulator

	Real Hardware	HDL Software Simulator	Hardware Accelerator	ASIC Hardware Accelerator	Emulator
Speed	Real time	Up to 10 clocks/s	Up to 100,000 clocks/s	Up to 500,000 clocks/s	Up to 1,000,000 clocks/s
Visibility	Poor	Excellent	Excellent	Excellent	Excellent
Compile time	Poor	Fast	Fast	Fast	Slow
Debugging tools	Poor	Excellent	Excellent	Excellent	Good
Checkpoint	No	Yes	Yes	Yes	Yes
Cost	Very high	Low	Medium to high	High	High
Testbench required	No	Yes	Yes	Probably	Probably
Simulation coverage	High	Low	Low to medium	Low to medium	Low to medium

is very expensive, and the compile times are excessive. It is more likely that an emulator can be connected to the real hardware environment, but even at 1 million clocks per second, tests that need billions of clock cycles will still take a long time. The biggest impediment to the emulator is the excessive compile times, which greatly extend the debug cycle.

3.5 FPGA PROTOTYPING

An FPGA prototype uses multiple FPGA devices to implement the design. The design is automatically or manually partitioned into blocks. These blocks are mapped using standard FPGA design software in standard FPGA devices. This approach differs from that of the emulator in the fact that the designers usually build a custom board to interconnect the FPGA devices instead of using FPGA interconnect devices. The board is a separate design effort from the FPGA devices themselves. The board creates fixed connections between the FPGA devices, and the boards usually have some type of RAM and ROM devices on them for the FPGA devices to store their configurations and to use as workspace.

The Good News

Using wires on a custom board to interconnect FPGA devices creates very fast FPGA-to-FPGA connections. FPGA prototypes can run tens of megahertz, approaching or meeting the speed of the real device. This can allow the designer to connect the prototype into the real environment. The designer is able to test the prototype with huge amounts of data from the real environment, as opposed to limited, manually created test sets. The verification process can be dramatically sped up as a result.

FPGA prototypes are relatively inexpensive, depending on the size and number of FPGA devices used. A design that consumes 10 or more of the largest devices can become expensive to build, as the board complexity could become large.

FPGA prototypes typically use very few devices compared to the same system built with an emulator. The device usage of an emulator is typically very small because of pin limitations and the fact that the emulator needs visibility to every signal. With an FPGA prototype the designer has control over which signals are visible so that fewer pins are required and device usage increases. Increasing device usage usually increases system speed as more pieces of the design are on the same device.

The Bad News

The FPGA prototype has the highest speed, but this comes at a price. The first problem is partitioning the design. Because of a limited number of pins on an FPGA device, partitioning can become a big problem. Depending on where the different block boundaries occur and the size of the design, a partition cut might require hundreds or thousands of signals between two blocks. Given the fact that there are only a few hundred available pins on the entire FPGA, pins quickly become the limiting issue for partitions. Designers may have to slow the interfaces and use pin multiplexing as in emulators.

The second problem with an FPGA prototype is the lack of internal visibility. With so few pins available compared to the number of internal signals, the amount of internal visibility can be a very low percentage, making debugging difficult. Techniques such as wiring internal signals to pins are used, but with limited success.

One very useful debugging technique used in FPGA prototypes is to employ on-chip RAMs to capture signal values in real time, then transfer the data out of the device through the standard JTAG interface of the device. This technique does not use any more external pins, yet gives the designer internal visibility into the inner workings of the FPGA prototype running at full speed.

By comparing the FPGA prototype to the other techniques as shown in Table 3.5, it can be seen that the FPGA prototype provides high speed of operation, yet suffers from slow compile times. This technique is not overly expensive, but it can take a long time to get the prototype working.

3.6 SUMMARY

Up to this point we have discussed why verification is so important for an ASIC design, the typical verification process, and the typical

Comparison with FPGA Prototype

	Real Hardware	HDL Software Simulator	Hardware Accelerator	ASIC Hardware Accelerator	Emulator	FPGA Prototype
Speed	Real time	Up to 10 clocks/s	Up to 100,000 clocks/s	Up to 500,000 clocks/s	Up to 1,000,000 clocks/s	Up to 10,000,000 clocks/s
Visibility	Poor	Excellent	Excellent	Excellent	Excellent	Poor
Compile time	Poor	Fast	Fast	Fast	Slow	Slow
Debugging tools	Poor	Excellent	Excellent	Excellent	Good	Poor
Checkpoint	No	Yes	Yes	Yes	Yes	No
Cost	Very high	Low	Medium to high	High	High	Low to medium
Testbench required	No	Yes	Yes	Probably	Probably	Maybe
Simulation coverage	High	Low	Low to medium	Low to medium	Low to medium	Medium

techniques used to verify designs. As can be seen by the comparison tables in this chapter, all of these techniques trade off speed, cost, flexibility, and a number of other factors. Which method is used greatly depends on the goals of the design team and the target application.

Chapter 4 describes another technique that enables the designer to prove mathematically that a design matches a specification. This technique can be used to help shorten the debug cycle and get your design to market faster.

Introduction to Formal Techniques

In this chapter, we introduce basic concepts related to the area formal verification—and specifically to proving functional correctness of a *register transfer level* (RTL) model. In general, functional verification is a process of checking that the intent of the design was preserved in its implementation. Hence, this process requires two key components: a *specification* and an *implementation*. For example, the design intent is often specified in a natural language such as a requirements document. The first step in the design and verification process is to translate the design intent to an implementation (such as an RTL model) as well as a specification form useful for verification (e.g., a formal specification, a behavioral reference model, or even a self-checking testbench). Janick Bergeron, in *Writing Testbenches* [Bergeron 2003], conceptually describes the verification process as a reconvergence model, where the verification of the various design transformations is reconciled back to a common source through a second reconvergent path.

Figure 4.1 demonstrates this concept. The upper path in the reconvergence model represents the transformation of the design intent into an RTL implementation. Similarly, the lower path of the reconvergence model represents the transformation of the design intent into a form of specification used by the verification tool. As

FIGURE 4.1

Reconvergent Paths in the Verification Process

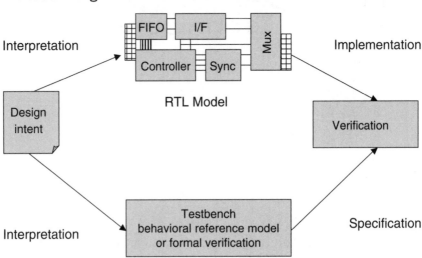

shown in the reconvergence model, the functional verification tool checks whether the design implementation conforms to its specification.

To check for conformance, we have two options:

♦ We can *demonstrate* that our design intent is preserved in our implementation, using dynamic approaches such as simulation.

♦ We can *prove* that our implementation satisfies the requirements of our specification, using formal verification.

Simulation to Demonstrate Conformance

To demonstrate conformance using simulation, we first create a testbench (or reference model), which captures the design requirements in a form suitable for simulation. In essence, the testbench becomes a refinement of our natural language specification during the verification process. Simulation stimulus is then generated into our implementation (e.g., our RTL model), and we monitor the implementation for incorrect behavior. When we use this simulation-based verification approach, our goal is to uncover the presence of a bug within our implementation. Once a bug is identified, our next step is to fix the problem and then reapply the same stimulus to our design to demonstrate that the bug no longer exists. We continue to reiterate this process until we are unable to identify any new bugs.

One drawback with this approach is that while simulation potentially demonstrates the presence of a bug (given the correct stimulus), it cannot ensure the absence of a bug. Another drawback is that simulation demonstrates one aspect (or effect) of a bug. Hence, the engineer often patches second- or third-order symptoms of the demonstrated bug, and not the actual root cause. Finally, validating the design intent through simulation is inherently incomplete for all but the smallest designs.

To illustrate this last point, let's attempt to exhaustively verify the functionality of a trivial 32-bit comparator, as shown in Fig. 4.2. Our specification for this example is quite simple:

- L is always a logical 1 whenever $A < B$.
- E is always a logical 1 whenever A==B.
- G is always a logical 1 whenever $A > B$.

To exhaustively verify our comparator example using simulation requires 2^{64} vectors. Let's assume our simulator is able to evaluate one vector every microsecond (which is incredibly fast). It would take 584,941 years to complete this verification process using simulation. In reality, verification engineers are smart, and they apply techniques to reduce the vector set (e.g., removing redundancy and focusing on selecting targeted corner cases). Nevertheless, today's designs are significantly more complicated than our simple 32-bit comparator example. Hence simulation, if

FIGURE 4.2

A 32-Bit Comparator

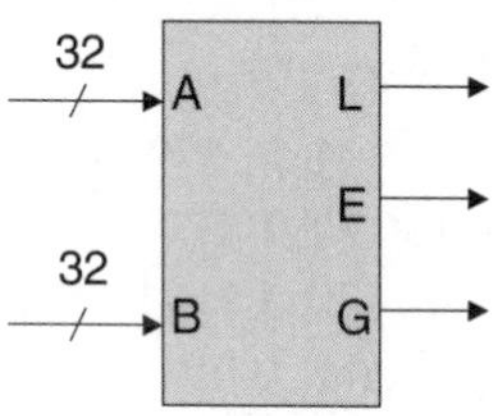

we are lucky, potentially identifies or demonstrates the *presence* of a bug; however, it cannot ensure the *absence* of a bug.

Formal Verification to Prove Conformance

Formal verification is a systematic process of ensuring, through exhaustive algorithmic techniques, that a design implementation satisfies the requirements of its specification. The keywords here are *systematic, specification,* and *exhaustive.* Note that functional formal verification, unlike simulation, does not use a testbench, simulation vectors, or any form of input stimulus to verify the design. This is so because the formal verification tool compiles the implementation model and specification into a mathematical representation, and then algorithmically and exhaustively proves that the implementation is valid with respect to the specification. To return to our 32-bit comparator example shown in Fig. 4.2, while it would take hundreds of thousands of years to completely simulate all 2^{64} vectors, functional formal verification can exhaustively verify this example within a matter of seconds without the need to create a testbench or any form of stimulus.

One key difference between a simulation-based methodology and a formal verification methodology lies in the approach to finding bugs. A simulation-based methodology requires the engineer to consider (and then figure out *how* to generate or hope that random simulation uncovers) all complex sequencing of events or intricate corner-case behaviors that might invalidate the design implementation. For example, the engineer in a simulation-based methodology must consider all unusual alignment of multiple unexpected input transactions to uncover a corner-case bug. Coverage techniques are employed to help identify holes in the input stimulus. In fact, a disproportional amount of focus within a simulation-based methodology is on *how* to create input stimulus versus to *what* they want it actually being checked.

In contrast, in a formal verification methodology, the engineer does not think about *how* to create input stimulus or corner-case scenarios required to break the design. That is, the engineer only needs to specify *what* she or he wants to prove (i.e., what to check), and then let the formal verification tools algorithms exhautively explore (without stimulus) the mathematical representation of the design to uncover all incorrect behaviors and corner-case bugs that would break the design. Often, due to the limited stimulus applied to the design, simulation uncovers only second- or third-order effects of a bug and not necessarily its root cause. Hence, the designer often patches a specific symptom without correcting the real problem. In contrast, since formal verification is exhaustive, it enables the engineer to be more productive by uncovering all violations to the specified requirement down to the root cause of the bug.

One of the objections often raised concerning formal verification is that if the user does not specify enough properties to check, then not all behaviors of the design will be verified. However, in reality, the same argument applies to simulation. If the testbench does not contain enough checkers on the output ports for the *design under verification* (DUV), then not all behaviors will be verified. For a more in-depth discussion related to creating a comprehensive specification, see Chap. 7, "The Formal Test Plan Process."

Observability versus Controllability

Fundamental to the discussion of functional verification is an understanding of the concepts of controllability and observability. *Controllability* is a measurement of the ability to activate, stimulate, or sensitize a specific point (i.e., a line of code or structure) within the design. Note that while in theory a simulation testbench has high controllability of its input ports for the design under verification, testbenches generally have poor controllability over internal points.

Observability, in contrast, is a measurement of the ability to observe the effects of a specific, internal, stimulated point. Thus, a simulation testbench generally offers limited observability since the testbench generally only observes the output ports of the DUV (i.e., often internal signals and structures are hidden from the testbench). To identify a design error using a simulation-based approach, the following conditions must hold:

- Proper input stimulus must be generated to activate (i.e., sensitize) a bug at some point in the design.
- Proper input stimulus must be generated to propagate all effects resulting from the bug to an output port.

It is possible, however, to set up a condition where the simulation input stimulus activates a design error that does not propagate to an observable output port. In these cases, the first condition cited above applies; however, the second condition is absent. Note that addressing the observability challenge within simulation-based methodology is analogous to solving an *automatic test pattern generation* (ATPG) problem. That is, ATPG consists of the following steps (see Fig. 4.3):

1. Enumerate a fault.
2. Figure out how to create a set of vectors to activate or sensitize the fault (i.e., justify the fault).
3. Figure out how to create a set of vectors to propagate the fault to an observable point.

If the simulation test is not set up correctly, the engineer can actually stimulate a line of code containing an error (which can be checked with a code coverage tool) but not observe the error. In fact, cases have been studied where the engineer achieved high-line coverage (on the order of 90 percent), yet only 54 percent of these lines would have been observed during the simulation process [Fallah et al. 1998]. As a consequence, bugs are missed if we only observe output ports during simulation.

FIGURE 4.3

ATPG Steps

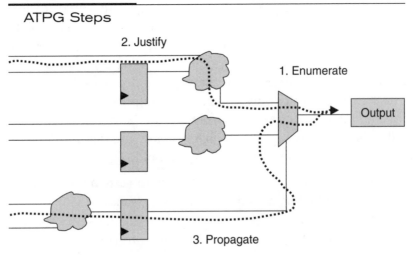

One means of increasing observability in a simulation-based methodology is by embedding lower-level or structural implementation assertions directly in the RTL model. In this way, the simulation environment no longer depends on generating a specific input stimulus to propagate the bug's effect to an observable point. Any improper or unexpected behavior can be caught by the assertion closer to the source of the bug, in terms of both time and location in the implementation, significantly reducing required debug time [Foster et al. 2004a]. However, it is important to note that while assertions improve the design's observability in a simulation-based methodology, it is still necessary to solve the controllability challenge to ensure a successful verification process. Hence, formal verification helps us overcome both the observability and the controllability challenge by algorithmically and exhaustively exploring all possible sequences of state within our DUV (i.e., it is unnecessary to figure out how to stimulate the design to achieve high observability).

4.1 FORMAL VERIFICATION CONCEPTS

Although we have decided not to focus on theoretical aspects of formal verification in this book—thus concentrating on practical applications of formal verification—we feel that an overview of the key concepts is important to effectively apply formal verification to your flow. Hence, this section introduces key formal verification concepts. Building on this foundation, we then introduce the language of formal verification by defining many often-used terms in the industry as well as throughout the remainder of this book.

4.1.1 What Is Formal Verification?

Formal verification is a systematic process of ensuring, through exhaustive algorithmic techniques, that a design *implementation satisfies the requirements of its specification*. By using a formal verification tool, all possible executions of the design are mathematically analyzed without the need to develop simulation input stimulus or tests. The main deficiency of formal verification is its limited capacity compared to simulation. In Chap. 8, "Techniques for Proving Properties," we will discuss techniques that have been effectively used to address some of the capacity limitations of formal technology.

4.1.2 Formal Boolean Verification

Transforming a high-level model representation of a design into a physical implementation involves many steps. Proving equality between these various design representations has historically been a challenge. One traditional approach to overcome this challenge involved abandoning the register transfer level model in favor of a single representation, i.e., a gate-level model. Consequently, the gate-level representation became the golden model of the design and was used for functional verification, timing analysis, and other forms of physical verification. However, this approach failed to clearly distinguish between functional and physical verification, thus impeding the overall verification flow.

In an alternative approach, the design team maintains the RTL representation as the golden model during functional verification. The team establishes equality by running regression simulations on both the RTL and gate-level models and comparing the output results. This alternative presents its own challenges, though, because using simulation to prove equivalence is incomplete. In addition, both approaches severely hinder a project's time-to-market goals by creating a simulation bottleneck in the design flow. These problems precipitated the search for improved equivalence-checking methods.

Today, mathematical reasoning techniques offer dramatic improvement over simulation and test vectors in establishing proof of equality for many designs. This form of reasoning, called formal verification, systematically ensures that a design's implementation (e.g., the revised model) satisfies its specification (the reference model). What traditionally took weeks and days to only partially check using a simulation-based approach can now be verified completely on many designs in a matter of hours and minutes using formal boolean equivalence verification. Furthermore, formal equivalence-checking tools are one of three key components of today's RTL static sign-off flow, along with static-timing-verifier and automatic test pattern generation tools.

Formal verification is revolutionizing the design process. Therefore, this section explores the area of formal boolean equivalence verification as applied to the RTL design flow. Although equivalence checking is known to be co-NP hard, multiple verification techniques combined with effective design practices can ensure success.

Equivalence-Checking Mechanics

To better understand the technology and mechanics employed in today's equivalence-checking tools, consider a typical computational equivalence model, as shown in Fig. 4.4.

We construct this model, called a *miter*, by creating a product machine between two *finite-state machines* (FSMs) M_1 and M_2. We form this product machine by

- Identifying all primary input and output pairs between the two FSMs
- Connecting each corresponding primary input pair together
- Connecting each corresponding primary output pair to an XOR gate

FIGURE 4.4

Product Machine

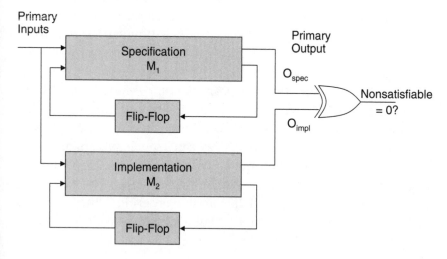

Proving equivalence between the two machines requires verifying that the product machine's XOR outputs are always 0 for any input sequence, once the machines have entered into their steady-state behavior.

Equivalence Proof Techniques

As a first step in understanding the general equivalence-checking process, consider the problem of proving combinational equivalence. Given two combinational circuits $F_{spec}(X)$ and $F_{impl}(X)$, where X represents an input vector defined on input variables $(x_1, x_2, \ldots, x_n)$, we can establish equivalence between the two circuits by proving that the following equation is nonsatisfiable for all values of X:

$$F_{spec}(X) \oplus F_{impl}(X)$$

Approaches developed to solve this problem include techniques based on *binary decision diagrams* (BDDs), ATPG, and simulations.

BDD-Based Techniques   Symbolic techniques, such as *reduced ordered BDDs* (ROBDDs, or simply BDDs), provide an efficient canonical-form representation for boolean functions, which can be used to prove equivalence. In many cases, we can easily verify equivalence between functions $F_{spec}(X)$ and $F_{impl}(X)$ by first constructing a BDD for each function, then determining if the two functions share the same BDD representation (canonical representations are isomorphic for equivalent functions). Alternatively, either the BDD for function $F_{spec}(X) \oplus F_{impl}X$ equals 0 when the two functions are equal, or the final BDD represents the set of all witnesses that distinguish the two functions.

Many practical functions can be represented with BDDs. However, a BDD's size is susceptible to the ordering of its supporting variables. Poor variable ordering can make memory size explode. In the worst case, BDD size can grow exponentially with respect to the number of inputs for certain functions. Nevertheless, designers can control the BDD size for many functions by exploiting the circuit structure during BDD variable ordering. Other methods include iterative improvement techniques based on BDD variable swapping.

In reality, solving the combinational equivalence problem is impractical when using a single global BDD to represent a circuit's output function. Efficient use of BDDs is possible, however, when the output function is partitioned into a set of subfunctions and then incrementally verified.

ATPG-Based Techniques   Another technique for proving equivalence, originally proposed by Roth [1977], uses ATPG. We check equivalence between two combinational circuits, such as the ones represented by functions $F_{spec}(X)$ and $F_{impl}(X)$, by running ATPG on a stuck-at-0 fault applied to the XOR product of the two functions $F_{spec}(X) \oplus F_{impl}(X)$ (see Fig. 4.4). If the stuck-at-0 fault is untestable, then the two circuits are equivalent. However, if the ATPG justification step identifies a witness for the stuck-at-0 fault, then the two circuits are unequal, and the ATPG distinguishing vector provides a debugging source.

Notice the differences between the BDD- and ATPG-based techniques. Whereas the complexity for BDD-based techniques resides in the space domain and potentially suffers from memory explosion, the complexity for the APTG-based techniques resides in the time domain and potential time-outs. Furthermore, BDD-based techniques identify all witnesses for the case when $F_{spec}(X)$ does not equal $F_{impl}(X)$, whereas ATPG-based techniques find only a single witness. Hence, if the two functions are suspected to be unequal, then applying ATPG is generally more efficient than using BDDs. Finally, when there is very little structural similarity between the two circuits, BDD-based techniques are generally more efficient at proving equivalence.

Without an incremental approach, applying purely ATPG-based techniques will not achieve successful verification for today's complex designs (the same is true for a pure nonincremental or global BDD-based technique). However, incremental verification techniques can improve the performance of both approaches.

SAT-Based Techniques   Satisfiability (SAT) algorithms search a conjunctive normal form (CNF) boolean formula for assignments that satisfy each of its conjunctive clauses. Note that SAT-based techniques have characteristics in common with ATPG-based techniques when applied to equivalence checking. Essentially, the equivalence-checking tool represents the $F_{spec}(X) \oplus F_{impl}(X)$ equation as a CNF formula, and determines if it is satisfiable (thus, proving that the two circuits are not equivalent). The F_{spec} and F_{impl} formulas represent the equations for the fan-in cone of logic to a register, and the X represents the cut in the circuit (which could be an output of a register, an input to the design, or potentially subcircuit cut known as equivalence point).

Simulation-Based Techniques   Although proving equivalence using simulation is incomplete for large designs, some equivalence checkers use simulation to exhaustively verify trivial functions. Additionally, simulation-based techniques can be combined with BDD-, SAT-, and ATPG-based approaches to create a powerful proof engine. For example, Burch and Singhal created a tight integration of verification approaches, which combine simulation- and BDD-based techniques with boolean SAT techniques.

Reducing Complexity

By exploiting the structural similarity between two designs, we can partition the combinational equivalence-checking problem into a set of smaller, simpler problems. Simulation is one technique used to identify structural similarity between two designs. For example, consider simulating a large set of random vectors. While calculating a signature on each internal signal (based on its response to the input vector), we can identify a set of candidate equivalent signal pairs between the two designs. These equivalent pairs are called *cut points*.

This section explores just a few incremental techniques for reducing the equivalence-checking complexity. Additional techniques for reducing complexity, such as recursive learning or transformation techniques, are not covered here. However, Huang and Cheng [Huang 1998] provide a comprehensive survey of the various incremental approaches used for proving equivalence.

Exploiting Structural Similarity Using ATPG-Based Techniques   Brand introduced an incremental ATPG-based method, which reduces the combinational equivalence-checking problem's complexity by exploiting two circuits' structural similarities. Although today's equivalence-checking tools do not use Brand's algorithm in its pure form, discussion of some of its details can help explain ATPG-based and incremental techniques [Brand 1993]. The steps involved in this method are as follows:

1. Construct a miter for the combinational circuit.
2. Identify the set of candidate equivalent pairs between the two circuits, e.g., internal signal f from model M_1 and internal signal g from model M_2, using simulation or other techniques.
3. Topologically order the candidate signal pairs, starting at the circuit's primary inputs and moving toward the primary output.
4. Run ATPG on a stuck-at-0 fault applied to the XOR product for the candidate equivalent pair $(f = g)$.
5. If the stuck-at-0 fault is found untestable $(f \wedge g)$, then signal g from model M_2 can safely replace signal f in model M_1, forming a merge point.

The last two steps are repeated for all ordered candidate signal pairs. This merging technique incrementally decreases the miter's size, potentially reducing the equivalence-checking complexity for the remaining candidate equivalent pairs.

Exploiting Structural Similarity Using BDD-Based Techniques Matsunaga [1996] as well as Kuehlmann and Krohm [1997] proposed alternative incremental methods that prove the internal equivalent signal pairs as local functions using BDDs. The input supporting variables to these local functions' BDDs are bounded by a cut set through the circuit (e.g., merge points that are topologically closer to the circuit's primary inputs). However, a problem known as a *false negative* can occur when one is attempting to prove these local functions using a BDD-based approach. For example, if an input cut point to the local function is assigned an impossible value with respect to values assigned on the circuit's primary input, then the local function might prove false. Multiple techniques have been developed to resolve the false-negative problem. One simplified approach is to move the cut set frontier

FIGURE 4.5

False Negative on Cut Point $x_1 = 1$ and $x_2 = 1$

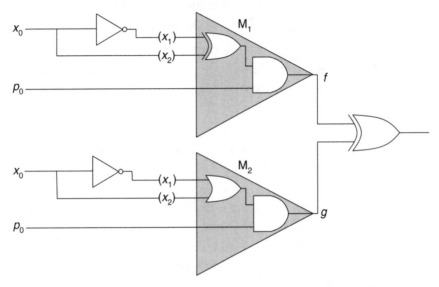

(related set of cut points) back toward the primary inputs, even though heuristics are generally used to optimize the cut point selection for resolution.

Figure 4.5 provides a simplified example of the false-negative problem. In this example, internal signals f, x_0, x_1, and x_2, and primary input p_0, which are all contained in model M_1, have corresponding cross-circuit equivalent pairs in model M_2. Hence, if we attempt to prove local function f with bounding cut points x_1 and x_2, then f proves false because of the impossible variable assignment $x_1 = 1$, $x_2 = 1$, and $p_0 = 1$. Yet, if local function f is viewed in the context of a larger logic cone, using cut point x_0 and primary input p_0, then f proves correct.

Equivalence Checking with Don't Cares

Thus far, we have discussed techniques for proving combinational equivalence between two models without don't cares. However, we need to consider additional issues when proving equivalence involving an RTL model. Typically, the RTL reference model's behavior is not fully specified for all external input values. This lets synthesis tools use the unspecified input space during the optimization steps. However, proving equivalence by using a reference model that is not fully specified can cause a false negative. Therefore, in don't-care cases, the computation model for combinational equivalence must be modified to account for the design's external don't-care input space. This can be represented mathematically, where DC_{spec} is the set of all input vectors contained in the specification model's don't-care input space. Then we can establish equivalence between the two models by proving the following equation is a tautology:

$$\forall X (\neg (F_{spec}(X)) \oplus F_{impl}(X))) \lor (X \in DC_{spec}))$$

In other words, either $F_{spec}(X)$ equals $F_{impl}(X)$ for all input values of X, or X is contained within the specification's don't-care input space. Furthermore, the DC_{spec} set can be extracted from the RTL model and represented as a gate-level network, or it can be represented symbolically as partitioned sets on the don't-care input space.

In addition to external don't cares, equivalence checkers need to account for an RTL model's internal don't cares, which result from using RTL synthesis directives (e.g., **full_case** or **parallel_case**) or explicit RTL 1'bX assignments (such as assign

```
c_control = (c_ready) ? c_value : 1'bX;
```

in Verilog). Both synthesis directives and explicit 1'bX assignments can create numerous verification challenges within an RTL static sign-off flow.

Sequential Equivalence Checking

The complexity of proving sequential equivalence is proportional to the design's state space. However, maintaining a consistent state space and state-encoding mapping between the specification and the implementation can reduce the complexity to a combinational equivalence-checking problem. Yet, in many cases, a one-to-one correspondence of the state space cannot be maintained. Examples include comparing a behavioral-level model to an RTL model or applying sequential transformations (such as retiming optimization) to a gate-level model. Hence, under these circumstances, sequential equivalence-checking techniques must be applied.

The process of proving sequential equivalence between two FSMs involves two steps. First, consistent presynchronization or transient behavior between the two machines must be established; then the machines' postsynchronization or steady-state behavior must be proved. Central to sequential equivalence checking is the definition of machine equivalence related to its transient behavior. In other words, do the two machines have a common reset state or a designated set of initial states, and do they have the same behavior after reset? Or, can an initialization or alignment sequence be applied to the two designs to bring them into an equivalent state, and are the machines' postsynchronization behaviors consistent (e.g., the notion of sequential hardware equivalence proposed by Pixley [1992])? Or, is the implementation circuit's I/O behavior contained within the specification circuit's I/O behavior during the alignment sequence, and are the machines' steady-state behaviors equivalent (e.g., the notion of safe replaceability proposed by Pixley [1994])? These questions hint at the details of the various notions of machine equivalence, which are beyond the scope of this book. However, Huang [1998] provides an excellent resource for the various equivalence definitions as well as equivalence-checking techniques in general (such as incremental techniques, register correspondence, debugging, and error localization).

To simplify the discussion on proving the steady-state behavior for sequential equivalence, assume that the two FSMs in Fig. 4.4 have a known equivalent reset state. The two machines are equivalent if the product machine's XOR outputs are always 0 for any reachable state. Hence, using FSM traversal techniques, we can check equivalence by computing the product machine's reachable states after reset. Obviously, this approach, like the general model-checking problem, encounters memory explosion for many of today's practical designs. Huang et al. [1997] have proposed a sequential equivalence-checking engine that combines techniques based on BDDs, advance sequential ATPG, and simulation to identify sequential similarities between the two FSMs. This approach, in an iterative manner, creates a reduced model for the sequential search.

Abstracting complex machines into simpler ones and applying symbolic simulation or using equivalence relations to reduce the problem are a few examples of new sequential techniques currently being explored. Although great strides are being made in sequential equivalence checking, much research remains before it can be integrated practically and successfully in today's large chip design flows.

RTL Static Sign-off

Using the concepts from this framework for proving equivalence, we now describe a static sign-off methodology. In a typical design flow, the completion of particular design stages requires sign-off approvals from either the design manager or the chip customer. Examples of these various sign-off stages include checks for functional correctness, timing, electrical design rules, RTL-to-gate equivalence, manufacturing test vector generation, and fault coverage analysis. Gate-level simulation historically has played a major role in verification as the tool used before many of these sign-off stages. Yet, as previously stated, using gate-level simulation to prove correctness provides low coverage and generally creates a bottleneck in the design flow.

As an example, in the late 1980s and early 1990s, the typical size of an ASIC design was about 50,000 to 100,000 gates. From the authors' own experiences during this period, verifying equivalence by means of running multiple gate-level regression simulations required approximately 3 to 4 weeks to complete, with unknown coverage results. To address these problems, many large companies (such as Hewlett-Packard, IBM, and Intel) began applying equivalence-checking research as an alternative static verification method, as well as developing disciplined approaches to RTL-based design. What had previously taken weeks to check using gate-level simulation can now be verified in minutes. The static sign-off methodology we are proposing included a high-performance equivalence checker, a static timing verifier, and an ATPG tool to achieve high manufacturing-fault coverage. This methodology enables our RTL model to truly remain the golden model of the design.

Although an RTL static sign-off methodology holds great potential for improving development time-to-market performance as well as increasing verification confidence, design engineers must closely examine several issues related to poor RTL coding practices before embracing this methodology. For instance, equivalence checkers can successfully verify that a design's boolean behavior is preserved during design transformations. However, they cannot verify that a model's simulation semantics have been preserved for the same transformations. This can result in pre- and postsynthesis simulation mismatches for logically equivalent circuits. What is particularly insidious are situations in which design errors cannot be demonstrated during RTL simulation yet are easily revealed using gate-level simulation. This inconsistency is due to RTL coding styles that mask a class of functional errors.

Figure 4.6 illustrates the RTL and gate-level models for a tristate-bus control circuit, which we have oversimplified for illustrative purposes. This example yields simulation mismatches between these two models. The RTL **full_case** synthesis directive asserts that all possible branches of a case statement are covered. This prevents the inference of a latch during the synthesis process while letting the synthesis optimizer treat unspecified branches as don't cares. However, unlike synthesis tools, simulators do not recognize the various synthesis directives. If a functional error exists in the RTL model that results in the illegal assignment of s=2'b11 (shown in Fig. 4.6a), then during simulation the en_n signal maintains its previous assignment; that is, en_n behaves as a latch. Because the previous latched value is valid, an RTL simulation might not detect the design error. In contrast, gate-level simulation quickly identifies the same functional error as a tristate-bus contention problem (as shown in Fig. 4.6b). We can avoid this problem by coding in a manner that preserves faithful semantics during synthesis.

To prevent mismatches between pre- and postsynthesis simulation, the simulator and synthesis tool must possess the same understanding of the RTL model [Bening and Foster 2001].

Examples of coding styles and tool directives that yield pre- and postsynthesis simulation mismatches, yet are logically equivalent, include

- **full_case** and **parallel_case** synthesis directives
- *X*-state assignment in the RTL

FIGURE 4.6

Semantic Inconsistency between RTL and Gate Model

```
wire (a,b,c);
reg (1:0)a;
reg (1:0)en_n;
..
..
always @(s)
begin
  case (s) //rtl_synthesis_full_case
    2'b00: en_n=2'b11;
    2'b01: en_n=2'b10;
    2'b10: en_n=2'b01;
  endcase
end
assign q=(-en_n(0)) ? a:1'bz;
assign q=(-en_n(1)) ? a:1'bz;
```

(a)   (b)

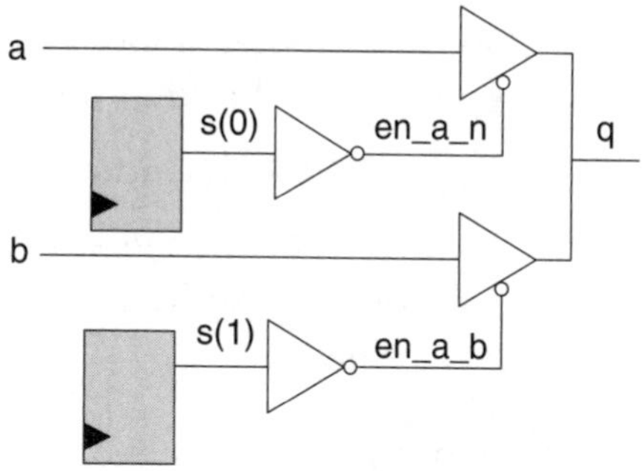

+ Initial blocks in the RTL (such as initial $q = 0;$)
+ An incomplete sensitivity list
+ Inadvertent latch inference (such as if without else, or if missing a default assignment)
+ Use of delays in the RTL (for example, #1, a Verilog delay construct)
+ Race conditions
+ Incorrect procedural-statement ordering

Emerging commercial tools combine traditional RTL linting with formal techniques to help the designer identify potential unfaithful-semantics problems. Because equivalence checking is a key sign-off step, ensuring faithful semantics is a necessary ingredient of a successful RTL static sign-off methodology.

Effective Equivalence-Checking Methodology

Today, equivalence checking is successfully applied to very large designs in industry. Obviously, the equivalence-checking tool's performance can vary depending on many factors, such as the logic cone's combinational depth, the size and number of multipliers, and the equivalence-checking technology. However, a poor design methodology can sabotage a good equivalence-checking tool. Here, we discuss methodology techniques that ensure optimal benefits during the equivalence-checking process.

To benefit from the emerging technology, we need to successfully manage tool integration. In an effective static sign-off methodology, equivalence checking must be

 ♦ Performed early and often throughout the design process
 ♦ Kept in the hands of the design engineer
 ♦ Combined with RTL recoding practices to optimize tool performance

Early and Often   Performing equivalence checking early and often throughout development helps resolve process issues before the sign-off stage. First, designers can resolve name-mapping issues before sign-off. Earlier, we introduced the idea of exploiting structural similarity to reduce equivalence-checking complexity. Although many commercial equivalence checkers will use name-mapping techniques to identify structural similarity between designs, these algorithms are not perfect. User intervention is often required to resolve mapping issues. To wait until the sign-off stage of design to resolve these issues is to invite trouble.

Second, performing equivalence checking early in the design process can help identify complex logic cones while there is still time to restructure the RTL to improve the equivalence checker's overall performance.

Finally, early equivalence checking can identify logic portions that might require alternative solutions for verification, such as logic cones that time out or complex multipliers.

In the Hands of the Design Engineer   In a traditional sign-off methodology, a project's design verification group performed equivalence checking using gate-level simulation. When commercial equivalence-checking tools became available, these verification groups continued to be responsible for the equivalence-checking process. The problem with this usage model occurs when the verification group detects an equivalence error. At this point, the verification engineer must consult the design engineer to debug the problem. This process can result in a bottleneck in the design flow. Furthermore, verification groups typically schedule the equivalence check as a single step in the flow rather than integrating the technology into the design process. A more effective methodology is to create a tighter, seamless integration of equivalence checking in the design process, as opposed to manually performing a single check within a verification process. We can accomplish this by including the equivalence check as an automatic step within multiple processes in the design flow.

RTL Recoding to Optimize Tool Performance

In an RTL-centric design flow, the RTL should be optimized to cooperate with the various tools in the flow. For example, we have found that profiling a large system simulation can result in simulation performance improvements on the order of 10 times. Although most performance issues are related to simulation logging or *programmable language interface* (PLI) C function calls, profiling quickly identifies poorly performing RTL code. In many cases, this code can be modified to improve simulation performance before regression testing reaches its peak. Before the advent of equivalence-checking tools, tuning RTL code for modules that negatively affected simulation performance was rare. Today, RTL recoding is safe and easily verified using equivalence checking.

In addition to tuning the RTL for simulation performance, it is often necessary to modify the RTL to improve synthesis results. Design engineers frequently need to specify explicit vendor cell instances directly within the RTL model. This action secures the timing performance that is difficult to obtain using logic synthesis. To prevent the loss of the designer's functional intent and protect the RTL code from degenerating into a slower-simulating gate-level net list, we recommend coding the RTL as follows:

```
`ifdef SPECIFICATION
        <RTL behavior specification>
else
        <macro cell instance
        implementation>
  `endif
```

For example,

```
`ifdef SPECIFICATION
        // calculate parity on `in'
        assign perr `^in;
`else // IMPLEMENTATION
        wire t1, t2, t3;
        XOR3 u1 (t1, in[0], in[1],
            in[2]);
        XOR3 u2 (t2, in[3], in[4]
            in[5]);
        XOR3 u3 (t3, in[7], in[9],
            in[10]);
        XOR3 u4 (perr, t1, t2, t3);
`endif
```

The SPECIFICATION Verilog text macro is defined during simulation and undefined during synthesis. This combination preserves clarity in the RTL description, optimizes the RTL for simulation performance, and ensures a specific implementation during synthesis. At this point, equivalence checking is reintroduced automatically to verify consistency between the RTL behavior description and the macro cell instance implementation—simply a self-compare on the RTL module. The reference model and the revised model are contained within the same RTL source. During the equivalence checker's compilation process, we define the SPECIFICATION text macro for the reference model and undefine it for the revised model.

By applying the techniques described in this section, we can build an effective RTL equivalence-checking methodology. Achieving a high number of equivalence checks raises the level of confidence that the design will ultimately meet its sign-off objectives without unforeseen problems.

4.1.3 Formal Sequential Verification

In Sec. 4.1.2, we discussed formal boolean verification, which is used to prove combitorial equivalence. In this section, we introduce the basic elements of formal property checking and, in so doing, convey a sense of both its inherent power and its limitations. Obviously, it is not necessary to fully understand all the details that occur under the hood of a formal verification tool to successfully apply formal verification to your design. However, we included the material in this section for completeness and believe the more general the understanding of formal technology you possess, the more likely you are to succeed at finding a solution to a performance issue that may arise during a proof.

Three steps are generally required to perform sequential formal verification (e.g., model checking):

1. Compile a formal model of the design.
2. Create a precise and unambiguous specification.
3. Apply an automated and efficient proof algorithm.

Each of these steps is briefly discussed below.

Compile a Formal Model

In the first step of the formal property-checking process, we create a formal model of the design by compiling an unambiguous description of the model (usually a synthesizable subset of a hardware description language, such as a Verilog RTL model) into a form accepted by the property checker. For the purpose of our discussion, we consider hardware designs as a set of finite-state concurrent systems. For example, the value of the current state of the system can be determined at a particular time by examining all state elements of the system. The next state of the system can be computed as a function of the system's current state value and design input values. This function is called a *transition function*. In formal verification, we can conveniently represent a current state–next state pair as a transition relation of the system. For example, (s_i, s_{i+1}) is a transition relation, where s_i represents a current state of the system and s_{i+1} represents one next-state possibility directly reachable from s_i.

In a formal representation of a design, we use *state* to indicate any variable retaining its value over time. In that broader sense, a state can be considered to be the inputs of the design, the variables from the control path representing the finite-state machines of the design and the variables from the data path representing stored results from operations.

A *path* π at state s is an infinite sequence of states $\pi = s_0, s_1, s_2, \ldots$ which represents a forward progression of time and a succession of states defined by a set of state transitions (s_i, s_{i+1}). Note that a simulation trace is one example of a path. A set of paths represents the behavior of the system. Hence, one form of a formal model can be created by automatically compiling a synthesizable RTL model of the design into a state transition graph structure, referred to as a *Kripke structure* [Kripke 1963].

A Kripke structure M is a 4-tuple $M = (S, S_0, R, L)$, which consists of

- A finite set of states S.
- A set of initial states S_0, where $S_0 \subseteq S$.
- A transition relation $R \subseteq S \times S$, where for every state $S_i \in S$, there is a state $s_j \in S$ such that $(s_i, s_j) \in R$.
- $L: S \rightarrow 2^{AP}$, where L is a function that labels each state with a set of atomic propositions (AP) that are true at that particular state.

A Kripke structure models the design using a graph, where a node represents a state and an edge represents a transition between states. Atomic propositions map boolean variables (and their negation) into the formal model of a design, represented by S, S_0, and R. For any atomic proposition p, if $p \in AP$, we say p is true or holds in s. Similarly, if $p \notin AP$, we say p is false or does not hold in s. By analyzing when an atomic proposition holds in a state, we can verify if a property is true or false.

Create a Precise and Unambiguous Specification

In the next step of formal property checking, we specify properties as assertions of the design that we wish to verify. Informally, a property describes design intent. More formally, we define a *property* as follows:

> A collection of logical and temporal relationships between and among subordinate boolean expressions, sequential expressions, and other properties that in aggregate represent a set of behavior (i.e., a path).

We define a *safety property* as follows:

> A property that specifies an invariant over the states in a design. The invariant is not necessarily limited to a single cycle, but it is bounded in time. Loosely speaking, a safety property claims that something bad does not happen.

More formally, a safety property is a property for which any path violating the property has a finite prefix such that every extension of the prefix violates the property.

For example, the property the signals wr_en and rd_en are mutually exclusive and whenever signal req is asserted, signal ack is asserted within 3 cycles are safety properties.

We define a *liveness property* as follows:

A property that specifies an eventuality that is unbounded in time. Loosely speaking, a liveness property claims that "something good" eventually happens. More formally, a liveness property is a property for which any finite path can be extended to a path satisfying the property.

For example, the property whenever signal req is asserted, signal ack is asserted sometime in the future is a liveness property.

Finally, we define a *fairness property* as follows:

A property that specifies that some condition will occur infinitely often. More formally, a fairness property is a property for which any infinite path will contain infinitely many fair states, or states that satisfy the fairness condition of the property.

Underlying many property languages is a formalism known as *propositional temporal logic*, which allows us to reason about sequences of transitions between states. Two formalisms for describing sequence propositions are *branching-time temporal logic* [Clarke and Emerson 1981] and *linear-time temporal logic* [Pnueli 1977]. CTL is an example of branching-time logic. The temporal operators of this formalism allow us to reason about *all* paths originating from a given state. Whereas in the case of LTL (a *linear-time temporal logic*), the temporal operators allow us to reason about events along a single computation path.

Apply a Proof Algorithm

Once we have created a formal model representing the design and a formal specification precisely describing a property that we wish to verify, the next step is to apply an automated proof algorithm. For example, given a formal model of a design described as a

FIGURE 4.7

Calculating Reachable States

fixed point: $S_k == S_{k+1}$

Kripke structure $M = (S, S_0, R, L)$ and a temporal logic formula f expressing some desired property of the design, the problem of proving the correctness of f involves finding the set of all states in S that satisfy f.

Note that the model satisfies the specification *if and only if* all initial states (that is, $\forall s_i \in S_0$) are in the set of the states that satisfies f (that is, we are interested only in paths in the Kripke structure that start from an initial state and satisfy f). One procedure for determining a set of states satisfying f is informally shown graphically in Fig. 4.7.

The illustrated proof algorithm we use is known as *reachability analysis using image computation*. This algorithm is the basic algorithm for proving that a temporal property f is valid.

The algorithm begins with a set of initial states S_0, as shown in Fig. 4.7. Using the transition relation R, as previously discussed, we calculate within one step (i.e., a tick of the clock) all reachable states from S_0. This calculation process is referred to as *image computation*. The new set of reachable states is S_1 in our example. We iterate on this process, generating a new set of reachable states at each step that grows monotonically, until no new reachable states can be added to the new set (i.e., a fixed point occurs when $S_k == S_{k+1}$). If the temporal formula f describes a safety property, we can validate that f holds on each new state calculated during the image computation step.

Proof Results

For this fixed-point proof algorithm, one of three possible results occurs:

- *Pass.* The process reaches a fixed point, and the formula f holds on all reachable states. Hence, we are done (i.e., the design is valid for this property).
- *Fail.* The process has yet to reach a fixed point, and the formula f was determined not to hold on a particular state s_j, which was calculated during the search. Hence, a counter-example (i.e., a path $\pi = s_0, s_1, s_2, \ldots, s_j$) can be calculated back from the bad state s_j to an initial state. This counter-example is then used to debug the problem.
- *Undecided.* The process aborts prior to reaching a fixed point due to a condition known as *state explosion* (i.e., there are too many states for the proof engine to represent in memory). In the following section, we discuss a few techniques that address the state explosion problem.

Formal property-checking tools use a number of different proof algorithms. A detailed discussion of these specific proof algorithms, creating formal models, and temporal logics is beyond the scope of this book. For in-depth discussions on these subjects, we suggest Clarke et al. [2000] and Kroph [1998].

4.2 SUMMARY

In this chapter, we introduced basic concepts related to the area formal verification—and specifically to proving functional correctness of a register transfer-level model. In general, functional verification is a process of checking that the intent of the design was preserved in its implementation. Hence, this process requires two key components: a *specification* and an *implementation*.

CHAPTER 5

Formal Basics and Definitions

The formal verification world tends to have its own language, which can often be confusing to novices as they enter this world and begin to explore formal methods. Hence, to assist the novice, we have created this chapter as a reference to commonly used terms and phrases from the formal world. We begin by briefly discussing a few basic algorithms commonly used by formal tools. Our focus is not to teach readers the details of these algorithms, but to familiarize them with the inherent power and limitations. A basic understanding of the various algorithms can also be useful when using some formal tools that permit the user to select various engines, as well as deciding what to do next when encountering tool or design complexity issues.

During PSL development, the Accellera Formal Verification Technical Committee [Accellera FVTC 2004] wanted to ensure that a common language would be used during our discussions. Hence, we identified a common set of terms and then developed definitions for these terms. In this chapter, we have expanded on that work, providing additional definitions related to other areas of formal verification (not limited to formal property languages). The reader might choose to skip this section and refer to it when needed.

5.1 FORMAL BASICS

For real-world formal verification applications, efficient algorithms than can manipulate boolean functions are required (e.g., determining if a boolean expression is satisfiable). Although there are often many different representations and algorithms used to manipulate boolean functions in use by today's formal verification engines (e.g., AIG, ATPG, BDD, SAT, Simulation), in this section our goal is to present a basic understanding of the inherent power and limitations of today's formal tools, due to their core algorithms. As we previously demonstrated in Chap. 4, to exhaustively verify a simple 32-bit comparator using simulation required 2^{64} vectors, and over 584,941 years to complete, while formal verification can exhaustively verify this example (using either BDDs or SAT algorithms) within a matter of seconds. However, with this inherent power also comes inherent limitations. That is, the core algorithms do not scale well for many real-world designs. Techniques to handle these limitations are discussed in Chap. 7.

5.1.1 Boolean Satisfiability

Given a propositional formula, the *boolean satisfiability problem* consists of determining if a variable assignment exists such that the formula evaluates *true*. This is often referred to as a SAT problem. Although SAT is an NP-complete problem, many large practical problems can be solved with today's SAT solvers [Zhang and Malik 2002]. One of the drawbacks with SAT algorithms is that the data structures do not provide canonical representation, which means that many subproblem evaluations are often repeated. Hence, SAT-based techniques are susceptible to time-out (i.e., for practical use they are time-limited).

Reduced ordered binary decision diagrams (or simply BDDs for our discussion) [Bryant 1986] are directed acyclic graphs providing a canonical representation of boolean formulas. For many instances, BDDs provide an exponential compression on the representation of a boolean formula.

The reason for the importance of BDDs in formal verification is that they provide canonical representation of boolean functions.

Canonical means that under certain conditions for every boolean formula, there is only one representation of this kind. This property is extremely important for verification since it enables us to prove equivalence between two formulas by first building BDDs (for each formula) and then checking to see if the two formulas are represented by the same BDD.

One of the drawbacks with BDD algorithms is that the size of the data structures is dependent on good variable orderings of the BDD nodes. The problem of finding an optimal variable ordering for many real-world applications is intractable. Hence many heuristics have been devised in order to derive acceptable variable ordering (static variable ordering problem) or to improve on an existing variable ordering (dynamic variable ordering problem). Hence, while SAT-based techniques are susceptible to time-outs, BDD-based techniques are susceptible to memory explosion (i.e., the circuit is too large to represent in memory).

The BDD variable ordering challenge is illustrated in the simple example of Fig. 5.1. There are some circuits, such as multipliers, for which the BDD representation will always reach exponential size with respect to the number of variables.

FIGURE 5.1

BDD Variable Ordering Challenge
[Boolean Function: a1(b&c)]

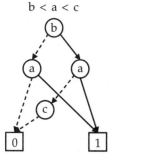
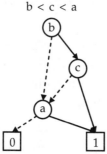

———————▶ Variable evaluates to 1
- - - - - - -▶ Variable evaluates to 0

FIGURE 5.2

BDD Application to Determine Equivalence

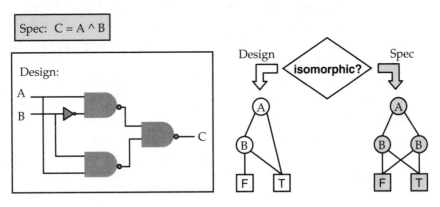

Figure 5.2 demonstrates an example of how BDDs can be applied during formal verification to determine if two different design representations are equivalent. The first step is to translate both designs to BDD representations and then to determine if the same single BDD was created representing equivalent circuits. For the example in Fig. 5.2, the design is not equivalent to its specification, since they do not share the same BDD (i.e., there are two different BDD representations, which means that the two circuits are not equivalent).

5.2 REACHABILITY ANALYSIS

BDDs can be used to efficiently represent a set of states [McMillian 1993]. This is important when one is proving properties on reactive systems, such as the system shown in Fig. 5.3.

Computing the set of reachable states using BDDs requires three fundamental components:

1. Representing sets of states using BDDs
2. Computing images, which is the set of all possible states that can be reached within one clock from a current state (or set of states)
3. Iterating, using images, to compute all reachable states

FIGURE 5.3

Modeling Finite-State Systems

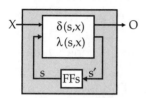

X, input; O, output; I, initial state; s, current state; s', next state; S, reachable state; λ(s,x), next state function; δ(s,x), output function.

For example, the next-state function $\lambda(s, x)$ describes a set of truth assignments and can be represented as a BDD. Furthermore, we can encode the states of the design, which in turn can be represented using BDDs.

Next, we can write a transition relation for the encoded states $R(S,S')$. For example, $R(s1,s2)$ means that state s1 transitions to s2.

Using the transition relation and a given state (or set of states), we can compute an image to determine within one clock all possible reachable states, or $\text{Img}(S, R)$.

Therefore, if we start our search at an initial state (e.g., a state right after reset) and continue to iteratively apply image computation on the new set of reachable states, we can continue to explore new sets of reachable states. For example, let R_0 be a BDD that represents the reset state. Then we can compute the set of reachable states as follows:

R_0 = BDD for reset state

$R_1 = R_0 \vee \text{Img}(R_0)$

....

$R_{i+1} = R_i \vee \text{Img}(R_i)$

This sequence will converge eventually, when $R_{i+1} = R_i$ (which is easy to test, since BDDs are canonical).

For an in-depth technical discussion on SAT and BDDs and their applications, we recommend Bryant [1986], Kroph [1998], and Zhang and Malik [2002].

5.3 DEFINITIONS

This section defines useful terms commonly found within the formal verification community. Many of the definitions were derived during the standardization effort of PSL by the Accellera Formal Verification Technical Committee [Accellera FVTC 2004]. We have expanded on that work, providing additional definitions related to other areas of formal verification (not limited to formal property languages). You will note that many of the definitions are focused toward the applied formal verification user.

assertion   A statement that a given property is required to hold and a directive to verification tools to verify that it does hold.

assumption   A statement that the design is constrained by the given property and a directive to verification tools to consider only paths on which the given property holds.

behavior   A path.

boolean   A boolean expression.

boolean expression   An expression that yields a logical value.

checker   An auxiliary process (usually constructed as a finite-state machine) that monitors simulation of a design and reports errors when asserted properties do not hold. A checker may be represented in the same HDL code as the design or in some other form that can be linked with a simulation of the design.

completes   A sequential expression (or property) completes at the last cycle of any design behavior described by that sequential expression (or property).

computation path   A succession of states of the design, such that the design can actually transition from each state on the path to its successor.

constraint   A condition (usually on the input signals) that limits the set of behaviors to be considered. A constraint may represent real requirements (e.g., clocking requirements) on the environment in which the design is used, or it may represent artificial limitations (e.g., mode settings) imposed to partition the verification task.

count   A number or range.

coverage   A measure of the occurrence of certain behavior during (typically dynamic) verification and, therefore, a measure of the completeness of the (dynamic) verification process.

cycle   An evaluation cycle.

describes   A boolean expression, sequential expression, or property describes the set of behaviors for which the boolean expression, sequential expression, or property holds.

design   A model of a piece of hardware, described in some hardware description language (HDL). A design typically involves a collection of inputs, outputs, state elements, and combinational functions that compute next state and outputs from current state and inputs.

design behavior   A computation path for a given design.

dynamic verification   A verification process in which a property is checked over individual, finite design behaviors that are typically obtained by dynamically exercising the design through a finite number of evaluation cycles. Generally, dynamic verification supports no inference about whether the property holds for a behavior over which the property has not yet been checked.

evaluation   The process of exercising a design by iteratively applying values to its inputs, computing its next state and output values, advancing time, and assigning to the state variables and outputs their next values.

evaluation cycle   One iteration of the evaluation process. At an evaluation cycle, the state of the design is recomputed (and may change).

extension   An extension of a path is a path that starts with precisely the succession of states in the given path.

fairness property   A property that specifies some condition will occur infinitely often. More formally, a fairness property is a property for which any infinite path will contain infinitely many fair states, or states that satisfy the fairness condition of the property.

false   An interpretation of certain values of certain data types in an HDL. In SystemVerilog and Verilog, the single-bit values 1'b0, 1'bx, and 1'bz are interpreted as the logical value false. In VHDL, the values STD.Standard.boolean'(False) and STD.Standard.Bit'('0'), as well as the values IEEE.std_logic_1164.std_logic'('0'), IEEE.std_logic_1164.std_logic'('X'), and IEEE.std_logic_1164.std_logic'('Z') are all interpreted as the logical value false.

finite range   A range with a finite high bound.

fix point   The fix point of a function f is any value x for which $f(x) = x$. For symbolic model checking, the termination state search process occurs when the computed new set of reachable states is the same as the last set of computed reachable states (that is, $S_k = S_{k+1}$).

formal specification   A concise description of design intent, often written in a language with mathmatically precise and well-defined semantics.

formal verification   A verification process in which analysis of a design and a property yields a logical inference about whether the property holds for all behaviors of the design. If a property is declared true by a formal verification tool, no simulation can show it to be false. If the property does not hold for all behaviors, then the formal verification process should provide a specific counterexample to the property, if possible.

holds   A term used to talk about the meaning of a boolean expression, sequential expression, or property. Loosely speaking, a boolean expression, sequential expression, or property holds in the first cycle of a path iff (if and only if) the path exhibits the behavior described by the boolean expression, sequential expression, or property.

holds tightly   A term used to talk about the meaning of a sequential expression. Sequential expressions are evaluated over finite paths (behavior). Loosely speaking, a sequential expression holds tightly along a finite path iff the path exhibits the behavior described by the sequential expression.

liveness property   A property that specifies an eventuality that is unbounded in time. Loosely speaking, a liveness property claims that "something good" eventually happens. More formally, a liveness property is a property for which any finite path can be extended to a path satisfying the property. For example, the property "whenever signal `req` is asserted, signal `ack` is asserted sometime in the future" is a liveness property.

logic type   An HDL data type which includes values that are interpreted as logical values. A logic type may include values that are not interpreted as logical values. Such a logic type usually represents a multivalued logic.

logical value   A value in the set {True, False}.

model checking   A method for formally verifying finite-state concurrent systems. Specifications about the system are expressed as temporal logic formulas, and efficient symbolic algorithms are used to traverse the model defined by the system and check if the specification holds.

monitor   *See* checker.

number   A nonnegative integer value, and a statically computable expression yielding such a value.

occurs, occurrence   A boolean expression is said to *occur* in a cycle if it holds in that cycle. For example, "the next occurrence of the boolean expression" refers to the next cycle in which the boolean expression holds.

path   A succession of states of the design, whether or not the design can actually transition from one state on the path to its successor.

path quantifier   A path quantifier is used to describe the branching structure in a computatonal tree, for example, $\forall$ ("for all computation paths") and $\exists$ ("for some computation paths").

positive count   A positive number or a positive range.

positive number   A number that is greater than 0.

positive range   A range with a low bound that is greater than 0.

prefix   A prefix of a given path is a path of which the given path is an extension.

property   A collection of logical and temporal relationships between and among subordinate boolean expressions, sequential expressions, and other properties that in aggregate represent a set of behaviors.

range   A series of consecutive numbers, from a low bound to a high bound, inclusive, such that the low bound is less than or equal to the high bound. In particular, this includes the case in which the low bound is equal to the high bound. Also, a pair of statically computable integer expressions specifying such a series of consecutive numbers, where one expression specifies the low bound of the series and the other expression specifies the high bound of the series. A range may describe time range, event repetitions, or bits of a bus or a vector. For time and repetition range, the low bound must be the left integer and the high bound must be the right integer. For vectors and bus bits range, the order is not important, unless restricted by the underlying flavor.

required (to hold)   A property is required to hold if the design is expected to exhibit behaviors that are within the set of behaviors described by the property.

restriction A statement that the design is constrained by the given sequential expression and a directive to verification tools to consider only paths on which the given sequential expression holds.

safety property A property that specifies an invariant over the states in a design. The invariant is not necessarily limited to a single cycle, but it is bounded in time. Loosely speaking, a safety property claims that "something bad" does not happen. More formally, a safety property is a property for which any path violating the property has a finite prefix such that every extension of the prefix violates the property. For example, the property whenever `signal req is asserted, signal ack is asserted within 3 cycles` is a safety property.

sequence A sequential expression.

sequential expression A finite series of terms that represent a set of behaviors.

simulation A type of dynamic verification.

starts A sequential expression starts at the first cycle of any behavior for which it holds. In addition, a sequential expression starts at the first cycle of any behavior that is the prefix of a behavior for which it holds. For example, if a holds at cycle 7 and b holds in every cycle from 8 onward, then the sequential expression `{a; b[*] ;c}` starts at cycle 7.

strictly before Before, and not in the same cycle as.

strong operator A temporal operator, the (nonnegated) use of which creates a liveness property.

temporal logic A formalism for describing sequences of transitions between states in a reactive system.

temporal operator A temporal operator describes properties of a path. There are five basic operators: X ("next time"), F ("eventually"), G ("always" or "globally"), U ("until"), R ("release").

terminating condition A boolean expression, the occurrence of which causes a property to complete.

terminating property A property that, when it holds, causes another property to complete.

transition A pair of states describing how the state of a system changes as the result of some action of the system.

transition relation If we think of s_i as the *present state* of a system and s_{i+1} as the *next state* of the system, then a set of pairs of states $R(s_i, s_{i+1})$ denotes a transition relation for the system.

true An interpretation of certain values of certain data types in an HDL. In the SystemVerilog and Verilog flavors, the single-bit value 1'b1 is interpreted as the logical value true. In the VHDL flavor, the values STD.Standard.boolean'(True), STD.Standard.Bit'('1'), and IEEE.std_logic_1164.std_logic'('1') are all interpreted as the logical value true.

verification The process of confirming that, for a given design and a given set of constraints, a property that is required to hold in that design actually does hold under those constraints.

weak operator   A temporal operator, the (nonnegated) use of which does not create a liveness property.

5.4 SUMMARY

In this chapter, we present a list of common terms often used within the formal verification community. In addition, we briefly discussed a few basic algorithms commonly used in formal tools. Our focus was not to teach the reader the details of these algorithms, but to familarize the reader with their inherent power and inherent limations.

CHAPTER 6

Property Specification

In this chapter, we discuss different approaches to solving the functional specification challenge. We begin by briefly introducing *propositional temporal logic*, which forms the basis for modern property specification languages. Building on this foundation, we then present an overview of emerging property specification languages and techniques for effective property specification for various classes of properties.

6.1 REASONING ABOUT CORRECT BEHAVIOR

Logic, whose origins can be traced to ancient Greek philosophers, is a branch of philosophy (and today mathematics) concerned with reasoning about behavior. In a classical logic system, we state a proposition and then deduce (or infer) whether a given model satisfies our proposition, as illustrated in Fig. 6.1.

For example, consider the following set of propositions:

+ The moon is a satellite of the earth.
+ The moon is rising (now).

FIGURE 6.1

Classic Logic System

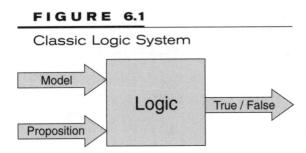

If we let the universe be our model, then using classical logic, we can check whether our set of propositions holds (i.e., evaluates true) for the given model.

Classical logic is good for describing static situations. However, classical logic is unsuitable for describing temporal behavior (i.e., situations involving time). To return to the previous example, it would not be possible to express the following proposition since it involves time:

• The moon will rise again and again.

Note that our interest in functional verification of hardware systems requires that we use a logic that is expressive enough to describe properties of reactive systems. For a reactive system, components of the system concurrently maintain ongoing interaction with their environment as well as with other concurrent components of the system. Hence, in the next section we discuss a more expressive logic that involves time.

6.2 ELEMENTS OF PROPERTY LANGUAGES

Today's emerging property language standards are based on a branch of logic known as temporal logic. The advantage of using temporal logic to specify properties of reactive systems is that it enables us to reason about these systems in a simple way. That is, temporal logic eliminates the need to explicitly describe time when specifying relationships between system events. For example, instead of writing the property expression as

∀t.!(grant1(t) & grant2(t))

(which states that for all values of time *t*, grant1 and grant2 are mutually exclusive), we write the property in a temporal language such as PSL (which implicitly describes time) as follows:

always!(grant1 & grant2)

(which states that grant1 and grant2 never hold or evaluate to true at the same time).

6.2.1   Linear-Time Temporal Logic

Propositional temporal logic is currently the most widely used specification formalism for reactive and concurrent systems. More specifically, linear-time temporal logic (for example, LTL [Pnueli 1977]), has emerged as the preferred foundational semantics for many of today's fledgling assertion property language standards (including PSL). A linear-time temporal logic strength is that it lets us reason about expected behavior over a linear sequence of states. This form of specification is intuitive to design engineers, since the specification thought process can view the progression of time similarly to a simulation trace. For example, using the PSL linear-time temporal operators, we can express a simple bus protocol property, that a transfer of type IDLE cannot be followed by type SEQ or BUSY as follows:

```
assert always (TRANS==`IDLE) ->
       next ((TRANS!=`SEQ) &&(TRANS!=`BUSY));
```

Given the temporal formulas f, f1, and f2, common PSL linear-time temporal operator examples include these:

!f	f does not hold
f1 & f2	f1 and f2 both hold
f1 | f2	f1 or f2 or both hold
f1 -> f2	f1 implies f2 (that is, !f1 | f2)
f1 -> f2	f1 -> f2 and f2 -> f1
always f	f holds in every cycle
never f	f does not hold in any cycle

next f	f holds in the next cycle
until f2	f1 holds until f2 holds
f1 **before** f2	f1 holds before f2 holds
eventually! f	f1 holds in some future cycle

Not all properties we are interested in can be expressed directly using linear-time temporal logic operators. For example, it is not possible to directly express a property that involves *modulo n* counting, such as

```
p is asserted in every even cycle
```

6.2.2 Extended Regular Expressions

Extended regular expressions overcome some of the limitations in linear-time temporal logic's expressibility. They are a convenient way to define a temporal pattern that can match (or, more aptly put, specify) sequences of states. Regular expressions let us describe design behavior that involves counting, such as modulo n type of behavior, with the * operator. Hence, regular expressions let us specify properties that cannot be described directly using linear-time temporal operators, such as p is asserted in every even cycle.

Sequences of boolean conditions that occur at successive clock cycles can be described succinctly using extended regular expressions. The advancement of time in a sequence is described by the ; operator in PSL (or ## in SVA). For example, the PSL extended regular expression

```
{a; b; [*3]; c; [*1:2]; d}
```

describes a set of sequences in which a occurs, followed 1 cycle later by b, followed 4 cycles later by c, followed between 2 or 3 cycles later by d. Obviously, this regular expression would match two unique sequences

a→b→true→true→true→c→true→d

and

a→b→true→true→true→c→true→true→d

where the arrow in this example illustrates a clock tick (ie., an advancement in time).

Both PSL and SVA support a form of extended regular expressions. Both support various repetition operators [ ] that concisely describe repeated concatenation of either a boolean expression or other sequences. For example, given the boolean expression b and sequence s, we can write

s[* n]   a sequence of n contiguous occurrences of s
b[= n]   any sequence containing n occurrences of b
b[-> n]   any sequence ending in the nth occurrence of b

The single constant n can be replaced by the repeat range m:n (for example, [*2:3]). PSL and SVA also support the following repetition operators:

s[*]   a sequence of 0 to infinity contiguous occurrences of s
s[+]   a sequence of 1 to infinity contiguous occurrences of s

Sequences can be composed with conjunctive operators as well as the time advancement operators. For example, given the sequences s1 and s2, in SVA we can create a new sequence by combining them as s1 ## s2, where the sequence s2 would start the clock immediately after s1.

To provide a real example, we could specify the following AMBA AHB forbidden sequence: When the master is not performing a burst type of INCR, then a transfer of type BUSY cannot be followed by NSEQ, which can be coded in SVA as follows:

```
property M_busy_and_not_incr;
  @(posedge HCLK) disable iff (~HRESETn)
    not ((HTRANS==`BUSY) & (HBURST!=`INCR)
      ## (HTRANS==`NSEQ));
endproperty
```

Even though regular expressions are a powerful way of describing sequences, they have their limitations; i.e., not all properties can be expressed directly using extended regular expressions. For example, the property eventually p holds forever cannot be directly expressed using extended regular expressions. However, it can be expressed using linear-time temporal logic constructs, e.g., `eventually always p`. Thus, linear-time temporal logic constructs address some of the limitations of extended regular expressions. Likewise, extended regular expressions complement linear-time temporal logic constructs.

6.2.3 High-Level Requirements Modeling

Although today's assertion languages, which combine linear-time temporal logic with extended regular expressions, are expressive, there are still many IP-related properties that cannot be directly expressed using this combination. For these properties, we can use a technique that combines linear-time temporal logic constructs, extended regular expressions, and modeling. For example, creating a complete specification for a PCI Express IP bridge requires specifying data integrity properties, such as a packet entering from the data link layer could never be dropped, duplicated, or corrupted as it progresses out to the physical layer [Loh et al. 2004]. Any property involving transaction ordering or queuing semantics requires storing information from various paths of data, which are then used during future data integrity checks. Hence, this type of specification often requires additional modeling to store the transaction information. Note that this process is similar to creating a scoreboard in a simulation testbench. By combining additional modeling with an assertion language, we can specify complex higher-level requirements that would not be possible by using only an assertion language's temporal constructs combined with extended regular expressions.

There are other instances when additional modeling combined with an assertion language simplifies the process of creating verification IP. For example, for some interfaces, such as the Intel PC SDRAM, the textual specification provides a command truth table that defines interface behavior. This form of defining behavior generally lists conceptual states of the interface, followed by valid commands (allowed with the current conceptual state) and the resulting new conceptual states due to the valid commands (i.e., the conceptual state transitions). Although extended regular expressions allow us to describe behavior in terms of sequences, attempting to describe behaviors that involve deep sequences that can further split into multiple sequences (describing alternate paths) can become unmanageable and result in less-than-obvious specifications. Since one of the goals of specification is to define behavior to eliminate ambiguities, specifications that are not clear are not desirable—nor easily maintainable. Hence, there are many instances where modeling aspects of the requirement both clarify and simplify the specification process.

6.3 PROPERTY LANGUAGE LAYERS

PSL is a comprehensive language that includes both *linear-time temporal logic* (LTL) and *computation tree logic* (CTL) constructs. The temporal semantics of PSL are formally defined in the LRM. This formal definition ensures precision; it is possible to understand precisely what a given PSL construct means and does not mean. The formal definition also enables careful reasoning about the interaction of PSL constructs, or about their equivalence or difference. The formal definition has enabled validation of the PSL semantic definition, either manually or through the use of automated reasoning [Gordon 2003].

However, for practical application, it is imperative that PSL statements be grounded in the domain to which they apply. This is accomplished in PSL by building on expressions in the hardware description language (HDL) used to describe the design that is the domain of interest. At the bottom level, PSL deals with boolean conditions that represent states of the design. By using HDL expressions to represent those boolean conditions, PSL temporal semantics are connected to, and smoothly extend, the semantics of the underlying HDL.

Before we introduce various forms for expressing assertions, it is helpful to consider definitions for two fundamentals: *property* and *event*. The reader should focus on the concepts presented in this section and not any specific syntax used to express these ideas. Details related to various assertion language syntax and semantics are discussed near the end of this chapter as well as in the appendices.

When one is studying property languages, such as PSL, it is generally easier to view their composition as four distinct layers. The *boolean layer* is the foundation; it supports the *temporal layer*, which uses boolean expressions to specify behavior over time. The *verification layer* consists of directives that specify how tools should use temporal layer specifications to verify functionality of a design. Finally, the *modeling layer* defines the environment within which verification is to occur.

Defining (or partitioning) a property in terms of the abstract layer view enables us to dissect and discuss various aspects of properties. However, you will find that it is quite simple to express design properties. Thus, the four-layer view is merely a way to explain concepts and should not convey a sense that the actual language syntax is complex.

6.3.1 Boolean Layer

The boolean layer consists of expressions whose values map to *true* or *false*. These expressions represent conditions within the design (e.g., the state of the design, or the values of inputs, or a relationship among control signals). For example, if we state that `signal grant1` and `signal grant2` are mutually exclusive, then the boolean layer description representing this property could be expressed in Verilog as shown in Example 6.1.

Notice that we have not associated any time relationship to the statement `signal grant1` and `signal grant2` are mutually exclusive. In fact, the statement by itself is ambiguous. Is this statement true only at time 0 (as many formal tools infer), or is it true for all time?

6.3.2 Temporal Layer

The temporal layer allows us to specify time relationships, and boolean expressions that hold (i.e., evaluates *true*) at various times.

EXAMPLE 6.1

A Property's Boolean Layer Expressed in Verilog

```
!(grant1 & grant2) // multiple grants cannot occur
```

EXAMPLE 6.2

A Property's Temporal Layer Expressed in PSL

```
always (!(grant1 & grant2)) // multiple grants cannot occur
```

This includes expressions that hold pseudocontinuously as well as expressions that hold at selected times, such as those points at which a clock edge occurs or an enabling condition is *true*. Temporal operators enable the specification of complex behaviors that involve multiple conditions over time. Thus, all time ambiguities associated with a property are removed. For example, if `signal grant1` and `signal grant2 are always mutually exclusive` (i.e., for all time), then a temporal operator could be added to the boolean expression to state precisely this. Temporal operators allow us to specify precisely when the boolean expression must hold. Example 6.2 demonstrates this point using the PSL temporal operator `always` combined with a Verilog boolean expression.

PSL provides temporal operators that can be used to specify sequences using extended regular expressions. A sequence describes behavior in the form of a series of conditions that hold in succession. A sequence may specify that a given condition repeats for a minimum, maximum, or even unbounded number of times before the next condition holds; and it may specify that two or more subordinate sequences overlap or hold in parallel. If the behavior described by a sequence matches the behavior of the design starting at a given time, then the sequence holds at that time.

6.3.3 Verification Layer

While a property's boolean and temporal layers describe general behavior, they do not state how the property should be used during verification. In other words, should the property be asserted and thus checked? Or, should the property be assumed as a constraint? Or, should the property be used to specify an event used to gather functional coverage information? Hence, the third layer of a property, which is the *verification layer*, specifies how sequences and properties are expected to apply to the design, and therefore how verification tools should attempt to verify the design using those sequences and properties.

Consider the following definitions for an assertion and a constraint.

Assertion—a given property that is expected to hold within a specific design. The PSL assert directive would be associated with the property to specify an assertion.

Constraint—a condition (usually on the input signals) which limits the set of behaviors to be considered during verification. A constraint may represent real requirements (e.g., clocking requirements) on the environment in which the design is used, or it may represent artificial limitations (e.g., mode settings) imposed to partition the verification. In this case, the PSL assume or restrict directives would be associated with the property to specify a constraint.

6.3.4 Modeling Layer

The modeling layer consists of HDL code used to model the environment of the design. In addition, the modeling layer is used to build auxiliary state machines that simplify the construction of PSL sequences or properties. Modeling the environment of design under verification is primarily of interest for formal verification. Building auxiliary state machines applies to both formal verification and simulation.

6.3.5 Events

When one is discussing design properties in the context of verification (and, in particular, simulation), it is helpful to understand the concept of a verification event. An *event* is any user-specified property that is satisfied at a specific time during the course of verification. A *boolean event* occurs when a boolean expression evalu-

E X A M P L E   6.3

A PSL Functional Coverage Point

```
cover {request2; [0:2]; grant2};
```

ates *true* in relation to a specified sample clock. A *sequential event* is satisfied at the end of a sequence of boolean events.

In Example 6.3, if the sequence `request2` followed by a `grant2` within 1 to 3 cycles is satisfied during simulation, then we can claim that an event has occurred in our verification environment at that specific time. However, if the event is never satisfied, then our verification test was unable to verify some key aspect or functionality of our design. In other words, our testing and input stimulus were insufficient. The PSL **cover** directive permits the designer to designate the property as a *functional coverage point*.

6.4 PROPERTY CLASSIFICATION

Properties are often classified in the context of their *temporal* and *verification* layers. Furthermore, properties can be classified by their *evaluation* methods (i.e., concurrent or sequential activation). This section describes the various classifications of properties.

6.4.1 Safety versus Liveness

As previously defined, a property is a general behavioral attribute that is used to characterize a design. It is generally expressed in a format that enables us to reason about sequences of boolean expressions over time. Hence, a property is often classified by its temporal layer. This section defines the two property classifications that are based on the temporal layer: *safety* and *liveness*.

A *safety* property is also known as an *invariant*, which informally states that, for all time, nothing bad should happen. Thus, it is a property that must evaluate to *true* for all sample points of time. The sample point could be defined by either an explicit clock associated with the property or an inferred clock.

A *liveness* property specifies an eventuality that is unbounded in time. Loosely speaking, a liveness property claims that something good *eventually* happens. For example, the property whenever signal req is asserted, signal ack is asserted sometime in the future is a liveness property.

6.4.2 Constraint versus Assertion

In addition to safety and liveness, a property can be classified according to its verification layer as either a *constraint* or an *assertion*.

One example of a *constraint* is a property that specifies the range of allowable values (or sequences of values) permitted on an input. The design cannot be guaranteed to operate correctly if its input value (or sequence of values) violates a specified constraint.

Alternatively, a property that describes that the expected design output behavior must remain valid or *true* is an example of an *assertion*. For any permissible sequence of input values applied to a design (which means that all input constraints are satisfied), all assertions will evaluate to *true* if the design is functioning correctly.

The functions of constraints and assertions are dependent on the verification tool and environment. During *simulation*, both constraints and assertions can be treated as monitors (i.e., dynamic property checkers) that check for compliance. During *formal verification*, constraints bound the static formal search engine to the design's legal input space, while assertions are treated as targets (i.e., properties that must be proved) for formal analysis.

6.4.3 Declarative versus Procedural

A *declarative property* describes the expected behavior of the design independent of its RTL procedural details. Hence, it is not necessary to understand the procedural code to understand the required expected behavior. On the other hand, a *procedural property* describes the expected behavior of the design in the current context (or frame of reference) at a particular line within the procedural code. Hence, it is necessary to understand the details of the procedural code to fully understand the expected behavior. Expressing interface properties declaratively is generally more natural than expressing these properties procedurally, since interface requirements are typically independent of the details of the block's implementation. However, capturing the internal RTL implementation's design intent procedurally generally reduces the amount of extra code required to express these properties (particularly if the assertion is deeply nested with case and if statements).

A design model typically consists of a static, hierarchical structure in which primitive elements interact through a network of interconnections. The primitives may be built-in simple functions (e.g., gates) or larger, more complex procedural or algorithmic descriptions (e.g., VHDL processes or Verilog always procedural blocks). Within a procedural description, statements execute in sequence. However, within the design as a whole, the primitives and communication interact concurrently.

Just as the design model itself involves a collection of concurrent elements (represented in a *declarative* fashion) and sequential elements (represented in a *procedural* fashion), properties may also be represented as either declarative or procedural. Hence, a *declarative assertion* is a statement (outside of a procedural context) that is always active and is evaluated concurrently with other layers or primitives in the design. A *procedural assertion*, on the other hand, is a statement within the context of a process (e.g., a Verilog procedural block) that executes sequentially, in its turn, as the procedural code executes.

6.5 PSL BASICS

As previously described in the introduction to this chapter, the advantage of using temporal logic to specify properties of reactive systems is that it enables us to reason about these systems in a simple way. In this section, we introduce a few of the basic PSL temporal operators. PSL is a very rich and expressive language. Not all operators are covered in this section. However, the operators we discuss will enable you to express a large class of common properties found in today's reactive system designs.

Note that PSL supports specifying and reasoning about synchronous designs with a single clock or with multiple clocks, as well as about asynchronous designs. For simplicity, many of our PSL examples are coded under the assumption that the engineer had previously defined a default clock in PSL, for example,

```
default clock = (posedge clk);
```

Furthermore, many of our examples, for simplicity, do not take into account a reset condition. We recommend that you augment our examples with the PSL **abort**:

```
assert always (a^b) @(posedge clk) abort rst_n;
```

6.5.1   PSL always and never Properties

The most basic forms of temporal properties are simple invariant (safety) properties, as discussed in Sec. 6.4.1. Examples of safety properties include these:

+ It is never possible to overflow a specific FIFO.
+ It is never possible to read and write to the same memory address simultaneously.

For instance, the PSL property shown in Example 6.4 states that signals grant1 and grant2 should always be mutually exclusive. The use of the temporal operator **always** indicates that the boolean expression ! (grant1 & grant2) must hold (i.e., evaluate *true*) at every cycle. Hence, PSL temporal operators enable us to reason about sequences of cycles.

The **never** PSL temporal operator allows us to specify conditions that should never hold. Thus, to state that grant1 and

E X A M P L E   6.4

PSL always Mutually Exclusive

```
always !(grant1 & grant2)
```

E X A M P L E   6.5

PSL never Mutually Exclusive

```
never (grant1 & grant2)
```

grant2 are mutually exclusive, we could have specified the code shown in Example 6.5.

Note that PSL properties are a declarative form of specification, which is independent of any hardware description language (i.e., not procedurally embedded). In other words, at the boolean layer PSL supports Verilog, SystemVerilog, or VHDL boolean expressions. The basic syntax for expressing a PSL **never** assertion (related to a boolean expression) is shown in Example 6.6.

PSL allows you to define named property declarations with optional arguments, which facilitates property reuse. These parameterized properties can then be instantiated in multiple places in your design with unique argument values. A property can be referenced by its name.

For example, we could specify that a and b are mutually exclusive whenever reset_n is not active, as shown in Example 6.7. The **abort** clause allows you to specify a reset condition. If the abort boolean expression becomes *true* at any time during the evaluation

E X A M P L E   6.6

PSL assertion Syntax

```
assert never (<boolean expression>) [@<clock expression>];
```

E X A M P L E   6.7

PSL property Declaration Example

```
property mutex (boolean clk, reset, a, b) =
   always (!(a & b )) @(posedge clk) abort !reset_n;
```

E X A M P L E 6.8

PSL Assertion for mutex grant1 and grant2

```
property mutex (boolean clk, reset, a, b) =
   always (!(a & b )) @(posedge clk) abort !reset_n;

assert mutex(clk_a, master_rst_n, grant1, grant2);
```

of the assertion expression, then the property holds regardless of the assertion expression evaluation.

In Example 6.8, we have created a PSL assertion for a design property where grant1 cannot occur at the same time as grant2.

6.5.2 PSL next Properties

In Sec. 6.5.1, we discribed the PSL always and never temporal operators which allow us to specify a condition that holds or does not hold for all cycles. To be more specific about a given cycle requires the use of other PSL temporal operators. For example, the PSL next temporal operator advances time forward by 1 clock cycle. Thus, to express the property whenever signal req is asserted, signal ack is asserted the next cycle, we would write the code shown in Example 6.9.

Note the use of the PSL boolean implication operator ->. In math, the implication operator consists of an *antecedent* that implies a *consequence* (for example, A -> C, which reads A implies C). If the antecedent is *true*, then the consequence must be *true* for the implication to pass. If the antecedent is *false*, then the implication passes regardless of the value of the consequence.

E X A M P L E 6.9

PSL Assertion for mutex grant1 and grant2

```
always (req -> next ack)
```

EXAMPLE 6.10

PSL Multiple next

```
always (req -> next (next (next ack)))
```

EXAMPLE 6.11

PSL Multiple next Using Repetition Operator

```
always (req -> next[3] ack)
```

To continue our example, if the ack is expected to hold on the third cycle after the req, then the property would have to be coded in a more complicated form, as shown in Example 6.10.

Although the specification for multiple **next** cycles shown above is valid, PSL provides a more succinct mechanism that utilizes the repetition operator [i], where *i* is a constant value. As shown in Example 6.11, **next**[3] states that the operand is required to hold at the third **next** cycle (rather than at the very **next** cycle).

6.5.3 PSL eventually Properties

As we demonstrated in Sec. 6.5.2, the next operator enables us to specify properties that advance time forward exactly 1 cycle. PSL provides a temporal operator that allows us to advance time forward without specifying exactly when to stop using the **eventually!** operator. For example, the property whenever signal req is asserted, signal ack should be asserted sometime in the future could be expressed as in Example 6.12. The exclamation point !, which is part of the name of the **eventually!**

EXAMPLE 6.12

PSL eventually! Operator

```
always (req -> eventually! ack)
```

operator, indicates that it is a strong operator. Strong versus weak operators will be discussed next.

6.5.4 PSL until Properties

The PSL **until** operator provides another way to reason about a future time, while specifying a requirement on a boolean expression that must hold for the current cycles moving forward (i.e., until a *terminating property* holds). This property states that whenever signal req is asserted, then starting at the next cycle, signal req will be deasserted until signal ack is asserted. For Example 6.13, boolean expression (i.e., signal) ack is the terminating property.

The **until** operator is a *noninclusive* operator; i.e., it specifies the left operand holds up to, but not necessarily including, the cycle where the right operand terminating property holds. As such, the subproperty (!req until ack) specifies that req will be deasserted up to, but not including, the cycle where ack is asserted. Thus, if signal ack is asserted immediately after the cycle in which the signal req is asserted, then the deassertion of req is not required.

Alternatively, the until_ operator is an *inclusive* operator; i.e., it specifies the left operand holds up to and including the cycle where the right operand terminating property holds. Thus, if the req signal is required to be deasserted (that is, !req) at least 1 cycle after the initial req, then **until_** would be used to specify this property, as shown in Example 6.14. This property states that whenever signal req is asserted, then !req will be asserted during the next cycle (whether or not ack is asserted), and it will remain asserted through (and including) the cycle where ack is asserted.

EXAMPLE 6.13

PSL Noninclusive until Operator

```
always (req -> next (!req until ack))
```

EXAMPLE 6.14

PSL Inclusive until_ Operator

```
always (req -> next(!req until_ ack));
```

Weak versus Strong Operators

One additional note concerning the PSL eventually, until, and until_ operators: these are known as *weak* operators. A *weak* operator makes no requirements about the terminating condition, while a *strong* operator requires that the terminating condition eventually occur. For example, the ack signal in Example 6.14 is not required occur prior to the end of verification (e.g., at the end of simulation) for the weak until_ operator. The eventually!, until!, and until!_ are all strong operators.

6.5.5 PSL before Properties

The PSL before operator provides way to reason about past behavior related to a specific event. For instance, suppose that we have a single-cycle active high request signal called req, and we have the requirement that before we can make a second request, the first must be acknowledged (i.e., ack). We can express this in PSL as in Example 6.15. We need the **next** operator to advance time 1 cycle to ensure that we are referring to some future req, and not the one we have just seen.

6.5.6 PSL Sequences

The basic PSL temporal operators described in the previous sections (i.e., **always, never, next, eventually, until,** and **before**)

EXAMPLE 6.15

PSL Inclusive until Operator

```
always (req -> next(ack before req));
```

E X A M P L E 6.16

PSL Sequence Specified with the next Operator

```
always
   (req -> next(ack -> next(!halt -> (grant & next grant))))
```

can be combined to create complicated properties. However, writing such assertions is sometimes cumbersome, and reading (and understanding) complicated assertions can be equally difficult.

The property shown in Example 6.16 states that the following sequence must occur (see Fig. 6.2):

* If signal req is asserted
* And then in the next cycle, signal ack is asserted
* And then in the following cycle signal halt is not asserted
* Then, starting at that cycle, signal grant is asserted for two consecutive cycles

PSL provides an alternative way to reason about sequences of boolean expressions that is more concise and easier to read and write. It is based on an extension of regular expressions, called *sequential extended regular expressions*, or SEREs.

SEREs describe series of boolean events by specifying a sequence for which each boolean expression in the series must hold over

F I G U R E 6.2

Sequence

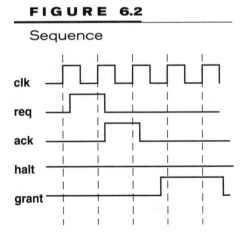

EXAMPLE 6.17

PSL SERE—Sequence of Boolean Expressions

```
{A&B; C|D; E^F}
```

contiguous cycles. A rudimentary SERE is a single boolean expression describing a boolean event at a single cycle of time. More complex sequences of boolean expressions can be constructed using the SERE concatenation operator ;. Example 6.17 shows the specification that three Verilog boolean expressions A&B, C|D, and E^F must hold consecutively.

The sequence is matched if the following three assertions hold:

♦ On the first cycle, the boolean expression A&B holds.

♦ On the second cycle, the boolean expression C|D holds.

♦ On the last cycle, the boolean expression E^F holds.

Often, an implication operator is used to start the sequence. Thus, if our specification states if signal req is asserted, then in the next cycle, signal ack must be asserted, and in the following cycle, signal halt must not be asserted, then the property could be written using a SERE as shown in Example 6.18. Note that if the req signal does not hold, then the sequences (defined by the SERE) that start in the cycle immediately after req are not required to hold.

Repetitions within Sequences

Like regular expressions found in most scripting languages (such as Perl or TCL), PSL allows the user to specify repetitions when specifying sequences of boolean expressions (SEREs). For instance, the SERE *consecutive repetition operator* [*m:n] describes repeated consecutive concatenation of the same boolean expression (or SERE)

EXAMPLE 6.18

PSL SERE for req, ack, !halt Sequence

```
always (req -> next {ack; !halt})
```

that is expected to hold between m and n cycles, where m and n are constants.

If neither of the range values is defined (that is, [*]), then the SERE is allowed to hold any number of cycles, including zero. Hence, the empty sequence is allowed. Also note that the repetition operator [+] is shorthand for the repletion [*1:inf], where the inf keyword means *infinity*.

For example, the SERE {a; b[*]; c[3:5]; d[+]; e} describes the following sequences:

- The boolean variable a holds on the first cycle of the sequence.
- And then, on the following cycle, there must be zero or more b's that must hold.
- This is followed by three to five c's that must hold.
- This is followed by one or more d's that must hold.
- This is finally followed by e that must hold.

Sequence Implication
In Example 6.18, we demonstrated a simple boolean implication in which a boolean expression implied a sequence. Often, it is desirable for the completion of one sequence (i.e., a prerequisite sequence) to imply either a property or another sequence. Hence, the *suffix implication* family of PSL operators enables us to specify this type of behavior.

The PSL *suffix implication* operator |-> can be read as follows: If the left-hand side prerequisite sequence holds, then the right-hand side sequence (or property) must hold. The | character symbolizes the completion of the prefix sequence, which is then followed by the implication operator ->, implying the suffix sequence.

Let us reconsider Example 6.16. Suppose that the 2 cycles of grant should start the cycle after !halt. We could code this as shown in Example 6.19. Or, we can simplify the code by using the repetition operator as shown in Example 6.20.

Note that the last boolean expression in the prerequisite sequence overlaps (occurs at the same time as) the first boolean expression in the suffix sequence. In other words, the !halt

E X A M P L E   6.19

PSL Suffix Implication

```
always ({req; ack; !halt} |-> {1; grant, grant})
```

E X A M P L E   6.20

PSL Repetition Operator

```
always ({req; ack; !halt} |-> {1; grant[2]})
```

E X A M P L E   6.21

PSL Suffix Implication with next Operator

```
always ({req; ack; !halt} |-> next {grant[2]})
```

E X A M P L E   6.22

PSL Suffix next Implication

```
always ({req; ack; !halt} |=> {grant[2]})
```

boolean expression in the prerequisite SERE overlaps with the first item in the suffix SERE. Hence, we add the 1 (*true*) boolean expression for this overlap, which moves time forward by 1 cycle. An alternative way to code Example 6.19 is shown in Example 6.21.

However, PSL provides a simpler way to do this using the *suffix next implication* operator | =>. The | => operator takes us forward in time by 1 clock cycle, which permits us to specify the property in Example 6.19 as shown in Example 6.22.

Declaring Sequences within PSL

In PSL, sequences can be declared and then reused with optional parameters.

For example, we could define a *request-acknowledge* sequence with parameters that allow redefining the req and ack variables as shown in Example 6.23. Once defined, a sequence can be reused and referenced by name within various PSL properties.

E X A M P L E   6.23

PSL sequence Declaration

```
sequence req_ack (req, ack) = {req; [*0:2]; ack};
```

E X A M P L E   6.24

PSL grant Overlapping Very Last req

```
{req[*1:4]}:{grant}
```

Sequence Operators within PSL

PSL provides a number of sequence operators, useful for composing sequences. In this section we introduce the *sequence fusion* operator and the *sequence length-matching AND* operation. The additional operators are described in App. A.

Sequence Fusion (:)   The *sequence fusion* operator : constructs a SERE in which two sequences overlap by 1 cycle. That is, the second sequence starts at the cycle in which the first sequence completes. For example, to specify that an active grant overlaps with the last active req in a set of sequences of one, two, three, or four consecutive req signals, we would code as in Example 6.24.

Sequence Length-Matching AND (&&)   The *sequence length-matching AND* operator (&&) constructs a SERE, in which two sequences both hold at the current cycle, and furthermore both complete in the same cycle.

6.5.7   PSL Built-in Functions

PSL contains a number of built-in functions that are useful for modeling complex behavior. In this section, we describe prev(), rose(), and fell(), which are used throughout various examples in the book.

   prev (bit_vector_expr [, number_of_ticks]):
   returns the previous value of the bit_vector_expr.

The number_of_ticks argument specifies the number of clock ticks used to retrieve the previous value of bit_vector_expr. If number_of_ticks is not specified, then it defaults to 1.

rose (boolean_expr): The built-in function rose() is similar to the posedge event control in Verilog. It takes a boolean signal as an argument and returns a *true* if the argument's value is 1 at the current cycle and 0 at the previous cycle, with respect to the clock of its context; otherwise, it is *false*.

fell (boolean_expr): The built-in function fell() is similar to negedge in Verilog. It takes a boolean signal as an argument and returns a *true* if the argument's value is 0 at the current cycle and 1 at the previous cycle, with respect to the clock of its context; otherwise, it is *false*.

6.5.8   PSL Verification Layer

In the previous sections, we demonstrated how to express properties using various PSL temporal operators. Yet, we never explicitly stated how these properties would be used during verification. The verification layer provides a set of directives that instruct a verification tool on what to do with these properties. In addition, it provides a way to group related directives and other PSL statements into verification IP (or verification units).

PSL provides a number of verification directives. The most basic verification directive is the assert statement, which instructs the verification tool to verify that the specified property holds (i.e., evaluates *true*). The verification directive assume instructs the verification tool to assume that a property holds, and is often used as a constraint on design inputs during verification.

Other verification directives work with SEREs rather than with PSL properties. For example, the verification directive restrict is a constraint that every trace must match a specified SERE. A restriction can be used to specify that an assertion should hold only for specific scenarios. For instance, a restriction can be used to specify a reset sequence. The verification direction cover directs the verification tool to check that a path matching the specified SERE was covered by a simulation test suite.

The PSL verification layer also provides verification units as a mechanism to group verification directives or other PSL statements, while specifying a name of the verification unit for reuse, and possibly binding information to a specific design module or module instance.

6.6 SYSTEMVERILOG ASSERTION BASICS

In this section we introduce the *SystemVerilog assertion* (SVA) operators [Accellera SystemVerilog-3.1a 2004]. Although, in general, the SVA operators are not as expressive as the full set of PSL temporal operators (i.e., SVA does not support all the LTL operators), it is still expressive enough for a large class of common properties typically embedded in an RTL design.

SystemVerilog recommends extensions to the IEEE-1364 Verilog language that permit the user to specify assertions declaratively (i.e., outside of any procedural context) or directly embedded within procedural code. In addition, SVA supports two forms of assertion specification: *immediate* and *concurrent*.

Immediate assertions evaluate using simulation *event-based semantics*, similar to other procedural block statements in Verilog. There is a danger of semantic inconsistency between the evaluation of immediate assertions in simulation versus formal property checkers, since formal tools generally evaluate assertions using *cycle-based semantics* (i.e., sampled off of a clock or signals) versus simulation event-based semantics. In addition, there is a risk of false firing associated with immediate assertions in simulation, which is discussed later.

Concurrent assertions are based on clock semantics and use sampled values of variables (note, this is similar to the OVL clock semantics). All timing glitches (real or artificial due to delay modeling and transient behavior within the simulator) are abstracted away.

For a detailed discussion of SVA scheduling and semantics related to assertion evaluation, we recommend Moorby et al. [2003].

6.6.1 SVA Immediate Assertions

Immediate assertions (also referred to as *continuous invariant assertions*) derive their name from the way they are evaluated in simulation. In a procedural context, the test of the assertion expression is evaluated *immediately*, instead of waiting until a sample clock occurs. When the variables in the assertion expression change values in the same simulation time slot, due to transient scheduling of events within a zero-delay simulation model, a *false firing* may occur if standard Verilog event scheduling is used. To prevent this class of false firings, evaluation of the assertion expression must wait until all potential value changes on the variables have completed (i.e., the transient behavior of events in the simulation has reached a steady state). Hence, SVA 3.1a has proposed a new region within the simulation scheduler's time slot called the *observe region*, which evaluates after the *nonblocking assignment* (NBA) region—ensuring that assertion expression variable values have reached a steady state [Moorby et al. 2003]. Note that this is similar to performing a PLI *read-only synchronization* callback to get to the end of the time slot region for safe evaluation. However, there is still a potential for false firings across multiple simulation time slots with immediate assertions, often due to a testbench driving stimulus into the DUV and delay modeling.

Syntax 6.1 defines SVA immediate assertions. Note that the SVA `assert` statement is similar to a Verilog `if` statement. For example,

SYNTAX 6.1

SVA Immediate Assertions

```
// See App. C for additional details.

immediate_assert_statement ::=
    assert ( expression ) action_block
action_block ::=
    statement    [ else statement_or_null ]
    | [statement_or_null] else statement_or_null
statement_or_null ::= statement    |  ';'
```

if the assertion expression evaluates to *true*, then an optional *pass statement* is executed. If the pass statement is omitted, then no action is taken when the assertion expression evaluates to *true*. Alternatively, if the assertion expression evaluates to 1'bx, 1'bz, or 0, then the assertion fails and the optional else *fail statement* is executed. If the optional fail statement is omitted, then a default error message is printed for whenever the assertion expression evaluates to *false*.

The optional assertion label (identifier and colon) associates a name with an assertion statement. And it can be displayed using the %m format code.

SVA has created a new set of system tasks (also referred to as a severity task) that are similar to the Verilog $display system task. These new tasks convey the severity level associated with an assertion's action_block while printing any user-defined message. The new severity tasks are as follows:

+ $fatal reports a runtime fatal severity level and terminates the simulation with an error code.

+ $error reports a runtime error condition and does not terminate the simulation. Note that if the optional fail state is omitted, the $error is the default severity level.

+ $warning reports a runtime warning severity level and can be suppressed in a tool-specific manner.

+ $info reports any general assertion information, carries no specific severity, and can be used to capture functional coverage information during runtime.

The details and syntax for these system tasks are described in App. C.

Example 6.25 demonstrates an SVA *immediate* assertion for our previous FIFO example.

6.6.2 SVA Concurrent Assertions

SVA concurrent assertions describe behavior that spans time. The evaluation model is based on a clock such that a concurrent assertion is evaluated only at the occurrence of a clock tick. SVA 3.1a has proposed a new region within the simulation scheduler's time slot,

EXAMPLE 6.25

SVA Queue Underflow Check

```
always @ (push or pop or cnt or reset_n)
  if (reset_n)
    if ({push, pop}==2'b01)
      underflow_check: assert (cnt!=0) else
                            $error("underflow error at %m");
```

called the *preponed region*, that evaluates at the beginning of a simulation time slot. Hence, the values of variables used in the concurrent assertion expression are sampled at the start of a simulation time slot, and then the concurrent assertion is evaluated using the preponed sampled values in the time slot observe region. Further details on concurrent assertion sampling are described in Moorby et al. [2003].

Property Declaration

As previously stated in this chapter, a *property* specifies a behavior of the design. Once defined, a property can be used in verification as an *assertion* (a property that is checked), a *functional coverage* specification (a property that must occur during verification), or a *constraint* (a property that limits the verification input space).

SVA allows you to define named property declarations with optional arguments, which facilitates property reuse. These parameterized properties can then be instantiated in multiple places in your design with unique argument values. A property can be referenced by its name. A hierarchical name can be used consistent with the SystemVerilog naming conventions.

For example, we could specify that a and b are mutually exclusive whenever reset_n is not active, as in Example 6.26. The

EXAMPLE 6.26

SVA property Declaration Example

```
property mutex (clk, reset_n, a, b);
    @(posedge clk) disable iff (reset_n) (!(a & b ));
endproperty
```

E X A M P L E   6.27

SVA property Declaration Example with not

```
property mutex_with_not (clk, reset_n, a, b);
    @(posedge clk) disable iff (reset_n) not (a & b);
endproperty
```

disable iff clause allows you to specify asynchronous resets. If the disable boolean expression becomes *true* anytime during the evaluation of the assertion expression, then the property holds regardless of the assertion expression evaluation. SVA also supports the specification of properties that must never hold, using the **not** construct. Effectively, the **not** construct negates the property expression. For example, we recode Example 6.26 as shown in Example 6.27. See App. B for specific details on SVA property syntax.

Verifying Concurrent Properties

After declaring a property, a verification directive assert or cover can be used to state how the property is to be used. The SVA verification directives include

- ◆ assert, which specifies that the property is to be used as an assertion (i.e., a property whose failure is reported during verification)
- ◆ cover, which specifies that the property is to be used as a functional coverage specification (i.e., a property whose occurrence is reported during verification)

In Example 6.28, we now create a concurrent assertion for a design property where write_en cannot occur at the same time as read_en.

Example 6.29 demonstrates an alternative form of directly specifying the same SVA concurrent assertion. Note that a concurrent assertion may be used directly within procedural code or alternatively stand alone as a declarative assertion within a module (i.e., outside of procedural code).

See App. B for specific details on SVA assert and cover syntax.

EXAMPLE 6.28

SVA Assertion for mutex write_en and read_en

```
property mutex (clk, reset_n, a, b);
    @(posedge clk) disable iff (reset_n) (!(a & b));
endproperty

assert_mutex: assert property (mutex(clk_a, master_rst_n,
                                          write_en, read_en));
```

EXAMPLE 6.29

SVA Simple Concurrent Assertions

```
assert_mutex: assert property @(posedge clk_a)
    disable iff (master_reset_n) (!(write_n & read_en));
```

SVA Sequences

A *sequence* is a finite series of boolean events, where each expression represents a linear progression of time. Thus, a sequence describes a specific behavior. An SVA *sequence expression*, like the PSL SERE previously discussed, describes sequences using *regular expressions*. This enables us to concisely specify a range of possibilities for when a boolean expression must hold.

Example 6.30 shows how we use SVA to concisely describe the sequence a request is followed 3 cycles later by an acknowledge.

In SVA, the ## construct is referred to as a *cycle delay* operator. The number after the ## construct represents the cycle in which the right-hand side boolean event must occur with respect to the left-hand boolean event. For the case ##0, both the left- and right-hand boolean events overlap in time (i.e., they occur in parallel).

EXAMPLE 6.30

SVA Sequence Expression with Fixed Delay

```
req ##3 ack
```

EXAMPLE 6.31

SVA Sequence Expression with a Range of Delays

```
req ##[2:3] ack
```

We can specify a time window with a cycle delay operation and a range. Example 6.31 uses SVA to describe the sequence a request is followed by an acknowledge within 2 to 3 cycles.

The previous examples are referred to as *binary delays*, or delays between two boolean expressions. SVA also permits us to specify *unary delays*, or boolean expressions that begin with a delay. Example 6.32 shows examples of unary delays.

Note that unary delays are useful when associated with implication. For example, if we want to describe a sequence in which a req must be followed by an ack within 2 to 3 cycles, which is then followed by a gnt, we write it as shown in Example 6.33 (using the SVA implication operator | ->).

Sequence Declaration In SVA, sequences can be declared and then reused with optional parameters, as shown in Syntax 6.2.

You can replace expression names within the sequence expression via parameters specified through the sequence_formal_list.

EXAMPLE 6.32

SVA Unary Delays Relationship to Binary Delays

```
(##0 start)       // that is, (start)
(##1 start)        // that is, (1'b1 ##1 start)
(##[1:2] start) // that is, (1'b1 ##1 start) or (1'b1 ##2 start)
```

EXAMPLE 6.33

SVA Unary Delays Relationship to Binary Delays

```
req |-> ##[2:3] ack ##3 gnt
```

SYNTAX 6.2

SVA Sequence

```
// See App. C for additional details.

sequence_declaration ::=
    sequence sequence_identifier [sequence_formal_list ] ';'
        { assertion_variable_declaration }
            sequence_expr ';'
    endsequence [ ':' sequence_identifier ]

sequence_formal_list ::=
            '(' formal_list_item { ',' formal_list_item } ')'
assertion_variable_declaration ::=
        data_type    list_of variable_identifiers
```

EXAMPLE 6.34

SVA Sequence Declaration

```
sequence req_ack (req, del, ack);
    req ##[1:3] ack; // ack occurs within 1 to 3 cycles after req
endsequence;
```

This enables us to declare sequences and reuse them in multiple properties. For example, we could define a request-acknowledge sequence
with parameters that allow redefining the req and ack variables as
shown in Example 6.34.

Sequence Operations   SVA defines a number of operations
that can be performed on sequences, such as these:

  • Specifying repetitions
  • Specifying the occurrence of two parallel sequences
  • Specifying optional sequence paths (e.g., split transactions)
  • Specifying conditions within a sequence (such as the
    occurrence of a sequence within another sequence or that
    a boolean expression must hold throughout a sequence)

◆ Specifying a first match of possible multiple matches of a sequence
◆ Detecting an endpoint for a sequence
◆ Specifying a conditional sequence through implication
◆ Manipulating data within a sequence

In this section, we focus on a set of common SVA sequence operators that we use in examples throughout the book. Details for all the SVA sequence operators are covered in App. B.

Repetition Operators   SVA allows the user to specify repetitions when defining sequences of boolean expressions. The repetition counts can be specified as either a range of constants or a single constant expression.

Like PSL, SVA supports three different types of repetition operators, as described next.

Consecutive Repetition

The consecutive repetition operator [*n:m] describes a sequence (or boolean expression) that is consecutively repeated with 1-cycle delay between the repetitions. Note that this is exactly like the PSL [*m:n] operator. For example,

```
expr[*2]
```

specifies that expr is to be repeated exactly two times. This is the same as specifying

```
expr ##1 expr
```

In addition to specifying a single repeat count for a repetition, SVA permits specifying a range of possibilities for a repetition.

SVA repeat count rules are summarized as follows:

◆ Each repeat count specifies a *minimum* and *maximum* number of occurrences, for example, [*n:m], where n is the minimum, m is the maximum, and $n \leq m$.
◆ The repeat count [*n] is the same as [*n:n].
◆ Sequences as a whole cannot be empty.
◆ If n is 0, then there must be either a prefix or a postfix term within the sequence specification.

♦ The keyword $ can be used as a maximum value within a repeat count to indicate the end of simulation. For formal verification tools, $ is interpreted as infinity (for example, [*1:$] describes a repetition of 1 to infinity). Note that this is similar to the PSL 1.0 inf keyword.

Nonconsecutive Count Repetitions

The nonconsecutive count repetition operator [=n:m] describes a sequence where one or more cycle delays are possible between the repetitions. The resulting sequence may precede beyond the last boolean expression occurrence in the repetition. Note that this is exactly like the PSL [=m:n] operator. For example,

    a ##1 b[=1] ##1 c

is equivalent to the sequence

    a ##1 !b [*0:$] ##1 b ##1 !b [*0:$] ##1 c

In other words, there can be any number of cycles between a and c as long as there is one b. In addition, there can be any number of cycles between a and the occurrence of b, and any number of cycles between b and the occurrence of c (that is, b is not required to precede c by exactly 1 cycle).

Note, the same sequence in PSL 1.0 would be coded as

    {a; b[=1]; c}

Nonconsecutive Exact Repetitions

The nonconsecutive exact repetition operator [->n:m] (also known as the *goto repetition* operator) describes a sequence where a boolean expression is repeated with one or more cycle delays between the repetitions and the resulting sequence terminates at the last boolean expression occurrence in the repetition. Note that this is exactly like the PSL 1.0 [->m:n] goto operator. For example,

    a ##1 b[->1] ##1 c

is equivalent to the sequence

    a ##1 !b [*0:$] ##1 b ##1 c

In other words, there can be any number of cycles between a and c as long as there is one b. In addition, b is required to precede c by exactly 1 cycle.

Note, the same sequence in PSL 1.0 would be coded as

```
{a; b[->1]; c}
```

The first_match Operator   The SVA `first_match` operator matches only the first occurrence of possibly multiple occurrences of a sequence expression. This allows you to discard all subsequent matches from consideration.

Syntax 6.3 describes the SVA `first_match` operator.

Consider an example with a variable delay specification as shown in Example 6.35. Each attempt of sequence `seq_1` can result in matches for up to four following sequences:

```
req ##2 ack
req ##3 ack
req ##4 ack
```

However, sequence `seq_2` can result in a match for only one of the above four sequences. Whichever of the above three sequences matches first becomes the result of sequence `seq_2`. Notice that

SYNTAX 6.3

SVA first_match Operator

```
// See App. C for additional details.

sequence_expr ::=
  first_match ( sequence_expr )
```

EXAMPLE 6.35

SVA first_match for req ack Sequence

```
sequence seq_1;
  req ##[2:4]ack;
endsequence

sequence seq_2;
  first_match(req ##[2:4]ack);
endsequence
```

SYNTAX 6.4

SVA throughout Operator

```
// See App. C for additional details.

sequence_expr::=
  expression_or_dist throughout sequence_expr
```

EXAMPLE 6.36

SVA Sequence with Boolean Condition

```
!interrupt throughout (req ##[2:4] ack #[1:2] gnt)
```

this is useful if the `ack` signal is held high for multiple cycles. The `first_match` prevents multiple unwanted matches from occurring.

The throughout Operators SVA provides a means for specifying that a specific boolean condition (i.e., an invariant) must hold throughout a sequence using the construct shown in Syntax 6.4. For example, to specify a sequence such that an `interrupt` must not occur during an `req-ack-gnt` transaction, we would code as in Example 6.36.

Dynamic Variables within Sequences SVA *dynamic variables* are local variables with respect to a sampling point within a sequence. The advantage of dynamic variables (over global variables) is that each time the sequence is entered, a new local variable is dynamically created. This ensures the sampling of data in overlapping sequence is correctly related to the appropriate sequence evaluation.

In Example 6.37 we demonstrate the usefulness of dynamic variables when validating the correct input/output data relationship in a pipeline register of depth 16.

Restrictions on dynamic variable usage, as well as syntax details, are defined in App. B.

E X A M P L E   6.37

SystemVerilog Dynamic Variable to
Validate Pipeline Latency

```
// pipeline register of depth 16
sequence pipe_operation;
  int x;
  write_en,(x = data_in)) |-> ##16 (data_out == x);
endsequence
```

SVA Implication Operators

The SVA implication operator supports sequence implication using the constructs in Syntax 6.5.

SVA provides two forms of implication: *overlapped* using operator `|->`, and *nonoverlapped* using operator `|=>`. The *overlapped implication operator* `|->` is similar to the PSL *suffix implication operator* `|->`, which can be read as If the left-hand side prerequisite sequence holds, then the right-hand side sequence must hold. Likewise, the *nonoverlapped implication operator* `|=>` is similar to the PSL *suffix next implication operator* `|=>`, which takes us forward in time by a single clock. For example, the nonoverlapped implication operator

```
(a |=> b)
```

is the same as the overlapped implication operator with a unary delay of 1:

```
(a |-> ##1 b)
```

S Y N T A X   6.5

SVA Implication Operators

```
// See App. C for additional details.

property_expr ::=
    sequence_expr |-> property_expr
  | sequence_expr |=> property_expr
```

The following points should be noted for sequential implication.

+ If the *antecedent sequence* (left-hand operand) does not succeed, implication succeeds vacuously by returning *true*.

+ For each successful match of the antecedent sequence, the *consequence sequence* (right-hand operand) is separately evaluated, beginning at the endpoint of the matched *antecedent sequence*.

+ All matches of *antecedent sequence* require a match of the *consequence sequence*.

6.6.3 SVA System Functions

SVA provides a number of new system functions useful for defining assertions:

+ **$past** (bit_vector_expr [, number_of_ticks], clock_enable, clock) returns a previous value of the bit_vector_expr. The number_of_ticks argument specifies the number clock ticks used to retrieve the previous value of bit_vector_expr. If number_of_ticks is not specified, then it defaults to 1. If the clock_enable is specified, the clock tick is counted when the expression is true. If the clock is specified, it is the clock for the evaluation. If it is not specified, the clock from the context of the expression is used.

+ **$isunknown** (<expression>) returns *true* if any bit of the expression is 'x' or 'z'. This is equivalent to

    ^<expression> === 'bx.

+ **$countones** (<expression>) returns a count that represents the number of bits in the expression set to 1. The 'x' and 'z' value of a bit is not counted toward the number of 1s.

See App. B for additional details related to SVA.

6.7   FAIR ARBITER EXAMPLE

Pointer-based round-robin arbitration is based on a circular moving pointer that identifies the specific port with the highest priority. For example, if port 0 was given a grant on the previous arbitration cycle, then port 1 will be designated the highest-priority port for the next arbitration cycle. However, if port 1 has no pending request during the current arbitration cycle, yet port 2 has a pending request, then port 2 will be issued the grant, which means that port 3 will be designated the highest-priority port and port 2 the lowest-priority port on the next arbitration cycle.

The advantage of pointer-based round-robin arbitration over other implementations is its simplicity. That is, after a grant is issued for a specific port, it is only necessary to change the pointer to identify the neighboring port as having the highest priority. The disadvantage of a pointer-based round-robin implementation is that if port 1 is a high-traffic port, port 0 will end up having the lowest priority most of the time. This is due to having continuous pending requests on port 1, while other ports have infrequent requests, which moves the pointer to the next-highest priority (port 2) and designates port 0 as a lower-priority port. This situation is usually not an issue, since at the time when port 0 issues a request, the fair arbitration scheme guarantees that it will receive a grant within N arbitration cycles, where N represents the number of requesting ports (i.e., clients).

For example, if the arbiter has eight clients, a pending request for a particular port should never have to wait more than 7 arbitration cycles before it is serviced. Otherwise, a different port must have been unfairly granted the bus multiple times.

The round-robin fairness property can be described as follows:

+ For any port request service, that particular port should receive a grant within N arbitration cycles, where N represents the number of requesting ports (i.e., clients).

Alternatively, the round-robin fairness property can be specified using a pair of ports (for example, p1 and p2) as follows:

+ If port p1 has a pending request, then port p2 should never receive two grants prior to port p1 receiving a grant.

Although p1 and p2 represent two fixed ports, we can use the technique from Sec. 8.4, "Symmetry," to cover all pairs of ports.

Property Name	Description
req_with_gnt	When a client generates a request and no other client has generated a request, then the next grant must be issued to the requesting agent.
gnt_no_req	No grant is issued without a request.
multiple_gnt	No more than one grant can be issued at a time.
Fairness	When a client generates a request, it should receive a grant within *N* arbitration cycles (i.e., within *N* issued grants).

For our example, req and gnt are vectors representing the current state of multiple requests and multiple grants for our system.

Although most arbiter interface requirements are relatively simple, there is often latency between the requests as they appear at the input of a block until they actually reach the arbiter circuit. Hence, for our examples, we introduce the Verilog macro definition (i.e., constant) 'LATENCY.

Note that p1 and p2 are two different pseudoconstants, which could be arbitrarily selected at start of verification and will not change value after selection (thus enabling us to check all possible ports in an arbitrary sense).

```
//  ****************************************************
// Single Request Grant Pair Requirement
//  ****************************************************
//  When a client generates a request, and no other
//  client has generated a request, then the next grant
//  must be issued to the requesting agent.
//  ****************************************************
req_with_gnt: assert always
        ({req[p1] && onehot(req)} |->
            {[*'LATENCY]; gnt[p1]}) @(posedge clock);
```

Note that a one-hot function is equivalent to ((req != 0 && (req & (req-1)) = 0)

```
//  ****************************************************
//  Single Grant
//  ****************************************************
//  This is a simple zero-one-hot check on gnt
//  Note that a one-hot check is (gnt & (gnt-1) == 0)
//  ****************************************************

single_gnt: assert always
        (gnt==0 || onehot(gnt));


//  ****************************************************
//  Formal Test Plan: Requirements Set 1: Grant without
//  Request
//  ****************************************************
//  No grant is issued without a request.
//  ****************************************************

gnt_wo_req: assert never
    {req[p1] == 1'b0; [*`LATENCY]; && (gnt[p1] == 1'b1)};


//  ****************************************************
//  Fairness Requirement
//  ****************************************************
//  APPROACH:
//  Monitor two ports at a time, verify one port at a
//  time, allow activity in all ports.
//  ****************************************************

// Property modeling layer

reg past_gnt;

 always @(posedge clk) begin
  if (req_delayed[p1]) begin
    if (gnt[p2]) past_gnt <= 1'b1; // seen a p2
    else if (gnt[p1]) past_gnt <= 1'b0; // a p1 occurred
  end
 end
fairness: assert never
      ({req[p1]; [*`LATENCY]} |=> {past_gnt & gnt[p2]})
           @(posedge clk);
```

6.8 SUMMARY

In this chapter, we discussed different approaches specifying design intent. We began by briefly introducing *propositional temporal logic*, which forms the basis for modern property specification languages. Building on this foundation, we then presented an overview of emerging property specification languages and techniques for effective property specification for various classes of properties.

CHAPTER 7

The Formal Test Plan Process

This chapter teaches you the steps necessary to create your own formal test plans. Our goal is to provide you with sufficient knowledge that you can create customized formal test plan chapters for your own design. In addition, we introduce a set of guidelines for effective assertion-based specification techniques. Following a good set of guidelines is crucial, for when engineers attempt to formally specify a design without proper guidance, it is not uncommon to generate a specification that is

- Complicated
- Difficult to comprehend through visual inspection
- Difficult to debug
- Not always as compact as expected

7.1 DEVELOPING A FORMAL TEST PLAN

One of the most challenging problems in creating a formal specification is to answer the questions

- Have I written enough properties?
- Is my property set complete?

This problem is not limited to a formal specification. In fact, this is an inherent verification problem. For example, similar questions are often asked concerning the comprehensiveness of the simulation testbench:

- ♦ Have I written enough checks in the testbench?
- ♦ Is my simulation environment complete?

Coverage metrics do not help us solve this problem since coverage measures the quality of the input stimulus and answers the question

- ♦ Are we missing a set of vectors required to reach a particular state in the design (or stimulate a line of code)?

Although current research in the area of determining the completeness of the property set exists, its focus is on identifying structures in the implementation that are not covered by an assertion (i.e., hole analysis) [Katz et al. 1999; Hoskote 1999; Chockler et al. 2001]. These approaches still fail to answer the question, "Is the property set complete?" With the lack of automated means to solve this inherent verification problem, we must depend on a systematic process to ensure high-quality formal specification. This section presents a set of steps that are useful as a process for creating a formal specification.

Step 1: Identify the Target Verification Approach

Your first step in developing a formal specification is to identify blocks for formal verification and blocks for simulation. Counter to what many believe, a single form of specification is not necessarily ideally efficient or effective for all verification processes. For example, it is generally more efficient in formal verification to specify a property on the output of a block with respect to a single input path through the block than to account for all input paths at once. When there are multiple input paths, it is recommended that you write multiple properties describing each path (input to output) separately.

The following discussion should help you decide which blocks are ideally suited for formal verification and which blocks are not.

FIGURE 7.1

Concurrent Paths

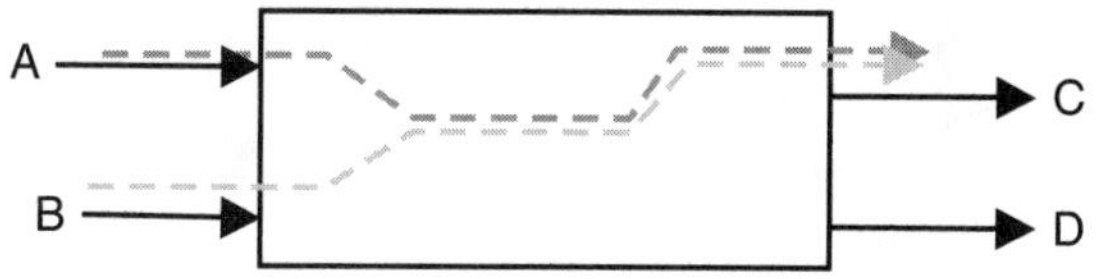

Blocks Suitable for Formal Verification   Formal verification is particularly effective for control logic and data transport
blocks containing high concurrency, as illustrated in Fig. 7.1.

The following list includes examples of blocks that we have
found ideally suited for formal verification:

- Arbiters of many different kinds
- On-chip bus bridge
- Power management unit
- DMA controller
- Host bus interface unit
- Scheduler, implementing multiple virtual channels for
  quality of service
- Clock disable unit (for mobile applications)
- Interrupt controller
- Memory controller
- Token generator
- Credit manager block
- Standard interface (e.g., PCI Express)
- Proprietary interfaces

This is an example of a bug identified by using formal verification
on a block involving concurrent paths:

> During the first 3 cycles of a transaction start from one side of the interface,
> a second transaction start unexpectedly came in on the other side of the
> interface and changed the configuration register.

Blocks Not Suitable for Formal Verification   In contrast, design blocks that generally do not lend themselves to formal

FIGURE 7.2

Sequential Paths

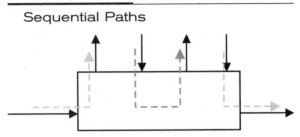

verification (i.e., model checking versus theorem proving) tend to be sequential (i.e., a single stream of data) and potentially involve some type of data transformation, as shown in Fig. 7.2.

Examples of designs that perform mathematical functions or involve some type of data transformation include

- Floating-point unit
- Graphics shading unit
- Convolution unit in a DSP chip
- MPEG decoder

An example of a bug associated with this class of design includes the following:

The IFFT result is incorrect if both inputs are negative.

Step 2: Describe Behavior

The second step in developing a formal specification is to briefly describe the behavior of the block to be verified. The importance of this step cannot be overstated. Without clearly articulating what it is you are planning to verify, your formal specification will be hopelessly incomplete. By describing and illustrating the intended behavior, you focus on what is to be verified. In fact, in many cases, errors in the design are identified through the thought process that occurs during this step—prior to any form of verification [Foster et al. 2004a].

This description does not have to be long or wordy. On the contrary, it can consist entirely of a set of block diagrams and interface waveforms that illustrate the expected end-to-end behavior accompanied by a brief description. It is, however, critical that you take this step prior to creating your assertion-based specification, to ensure completeness.

Step 3: Define the Formal Specification Interface

Identify the block's interface signals that must be monitored. Create a table that lists the signal names and their functionality. This list of signals will form your formal specification (e.g., high-level requirements module) interface. In addition to identifying the requirement's interface signals, this step describes any special considerations in the cycle timing relationships between the various interface signals that must be checked.

Step 4: Create Requirements Checklist

Create a requirements checklist, which is the list of requirements you plan to formally verify. We suggest that you create a table that lists the specific assertion name that you will code in your high-level requirements model (or verification IP), as shown in Fig. 7.3, along with a description in English of what the requirement checks. In a separate table, list all required assumptions.

When you are creating an assertion-based specification checklist, focus on what you want to verify versus how you plan to verify the design. This focus contrasts sharply with a traditional simulation-based test plan, which normally describes both what you want to verify (in terms of checks) and how you plan to verify the design (focusing on input sequences or scenarios as stimulus).

Often, you can derive the requirements checklist directly from an existing natural languages specification, such as a microarchitectural

FIGURE 7.3

Formal Specification
Requirements Model

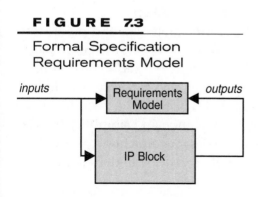

specification, a vendor's specification (e.g., the Intel PC SDRAM), or an industry protocol standard (for example, PCI). However, for some blocks (particularly a company's internal proprietary IP) there might not be (unfortunately) a clear specification. Hence, in those cases, you start the process of creating the requirements checklist in terms of the design outputs, not in terms of its inputs. Then you should think about end-to-end requirements versus internal structural or implementation requirements or assertions.

Step 5: Formalize the Natural Language Requirements Checklist

Using the interface signals identified in step 3 and the set of requirements identified in step 4, create your assertion-based specification. Encapsulate the set formalized requirements into a high-level requirements model that will monitor the IP's input and output signals, as shown in Fig. 7.3.

When you create your high-level requirements model concurrently during the design implementation phase (as opposed to prior to implementation), do the following:

1. Ask yourself, "Which high-level requirement from the requirements checklist should I specify first?"
2. Specify the ones that capture the essence of what the block is supposed to do from the microarchitecture perspective.
3. Specify the ones related to the most complete RTL functionality.

Step 6: Document Verification Strategy

In this step, you will define the verification plan and proof strategy. For example, you might choose to prove certain requirements by initially applying a set of restrictions (i.e., add constraints to restrict the explored behavior to only read-mode transaction, or only write-mode transactions). This approach lets you focus the verification on specific modes of operation and simplifies the verification process. Once the various modes function correctly, the restrictions are removed and the IP is verified under all modes of operation.

In addition, this step lists any verification strategies you might choose to apply during the verification process, such as abstraction or symmetry, for formal verification.

Step 7: Define Coverage Goals
The final step in the formal test plan process is to define a set of coverage goals that must be met during the verification process. For example, to ensure that you are verifying what you think you are verifying, check that constraints on specific states in your requirements model have not overconstrained the states. The checking procedure varies depending on your verification process. For example, in formal verification, you can check to see if specific states in your requirements model are reachable. For simulation-based approaches, you can apply functional coverage to various states or sequences of states in the requirements model.

7.2 RULES FOR WRITING A REQUIREMENTS MODEL

Although research shows there is little doubt that a formal verification is beneficial, early adopters have been forced to take a learn-as-you-go approach to writing formal properties (or what we refer to as a requirements model). To address the need for guidelines in this emerging field, this section, based on the experiences of early adopters and power users, lists a set of rules for effectively coding formal specifications.

Rule 1
Create a checklist of requirements prior to coding an assertion-based specification.

An ad hoc approach to assertion-based IP specification is likely to yield an incomplete, contradictory, and generally poor-quality specification. Follow the process outlined in Sec. 7.1 to ensure a high-quality complete specification.

Rule 2
Express all properties in a manner that is clear, concise, and obvious.

When you read the formal assertion-based specification, if the design intent is not obvious, then it is highly likely that the formal specification is wrong (or incomplete). Even if the specification is correct and complete, a complex assertion-based specification is not maintainable.

Do not be adamant about the specification style. For example, specifying properties solely with temporal logic or regular expressions is not possible. Use all means available. Strive for simplicity.

Guideline Partition large, complex properties into a set of simpler properties.

Guideline Partition large, complex sequences (specified using regular expressions) into a simpler set of sequence declarations.

Rule 3
Use an FSM model to describe conceptual states defined in the architecture.

Note: The behavior described in conceptual state machines should be intuitive. Once described, a simple assertion can then be specified over states contained in the conceptual state machine with respect to the IP design.

Guideline When the conceptual state machine becomes too large, try partitioning the FSM into a smaller set of FSMs. The partition should be driven by the properties we want to express. It is acceptable to describe overlapping behavior in the partitioned set of conceptual state machines since our goal is to provide just enough modeling to simplify assertion specification.

Rule 4
For converging data paths, specify the output behavior of a port with respect to a unique input path.

For example, in Fig. 7.4, to specify the correct behavior of port C with respect to its inputs, it is better to create a set of assertions. Create one assertion to describe the expected behavior of port C with respect to a specific data item on port A (allowing port B to assume any valid value that is not checked by the assertion). Then create

FIGURE 7.4

Converging Data Paths

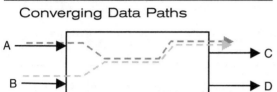

another assertion to describe the expected behavior of port C with respect to a specific data item on port B (again allowing port A to assume any valid value that is not checked by the assertion).

7.3 AMBA AHB EXAMPLE

In this section, we illustrate the process of developing a formal test plan using the AMBA AHB as an example. Note that it is not our intention to fully specify the AMBA AHB in this example.

7.3.1 Determine Target Verification

Your first step in developing a formal specification is to identify blocks for formal verification and blocks for simulation. The AMBA AHB interface is a good candidate for formal verification since it consists of a substantial amount of control logic and data transport logic containing high concurrency.

7.3.2 Describe Behavior

The second step in creating a developing a formal specification is to briefly describe the behavior of the block to be verified. AMBA is an on-chip communications standard originally intended for designing high-performance embedded microcontrollers for use with ARM processors. The AMBA specification is an established, open methodology that serves as a framework for SoC designs, effectively providing the "digital glue" that binds IP cores together.

The specification used in this discussion is the AMBA Specification (Rev. 2.0), which is copyrighted by ARM Ltd., 1999.

FIGURE 7.5

Conceptual Implementation of AMBA AHB and APB

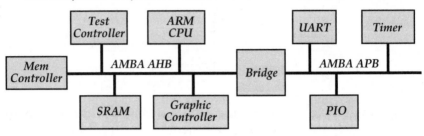

You can download the open AMBA specification free of charge from the ARM website; however, registration is required.

In Fig. 7.5, we illustrate a conceptual implementation of both the AMBA AHB and the AMBA Advanced Peripheral Bus (APB). In this chapter, we focus on creating a formal test plan and set of requirements for the AMBA AHB.

The AMBA AHB is designed to connect embedded processors such as an ARM core to high-performance peripherals, DMA controllers, on-chip memory, and interfaces. It is a high-speed, high-bandwidth bus that supports multimaster bus management to maximize system performance.

A typical AMBA AHB system design contains the following components:

- ◆ *AHB master.* A bus master is able to initiate read and write operations by providing address and control information. Only one bus master is allowed to actively use the bus at any one time.

- ◆ *AHB slave.* A bus slave responds to a read or write operation within a given address-space range. The bus slave signals back to the active master the success, failure, or waiting of the data transfer.

- ◆ *AHB arbiter.* The bus arbiter ensures that only one bus master at a time is allowed to initiate data transfers. Even though the arbitration protocol is fixed, any arbitration algorithm, such as highest priority or fair access, can be implemented depending on the application requirements.

FIGURE 7.6

AMBA AHB Master Interface

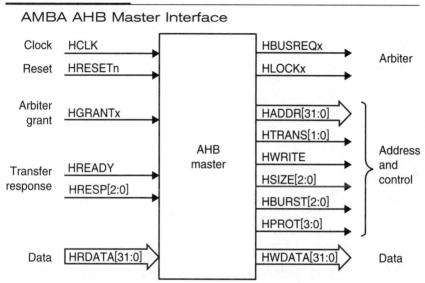

* *AHB decoder*. The AHB decoder decodes the address of each transfer and provides a select signal for the slave that is involved in the transfer. A single centralized decoder is required in all AHB implementations.

Figures 7.6 and 7.7 illustrate the interface signal connections for an AMBA AHB master and slave device.

For your own formal test plans, at this point you would create waveforms to describe the various interface transactions. For our example, the reader should reference the AMBA specification for those details.

7.3.3   Define Formal Specification Interface

The third step is to identify the block's interface signals that must be monitored. See Table 7.1.

7.3.4   Create Requirements Checklist

The fourth step is to create a checklist of requirements that must be formally verified. The can be viewed as the formal test plan.

FIGURE 7.7

AMBA AHB Slave Interface

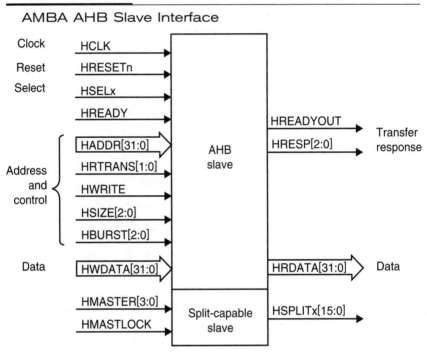

The AMBA AHB has over 80 requirements that would normally be present in the requirements checklist. However, our goal in this example is to illustrate the process. Hence, we will list only three of the requirements associated with HTRANS, and then we demonstrate how to formally specify these requirements.

Name	Summary
err_M_HTRANS_rst_value	HTRANS set to IDLE after reset.
err_M_HTRANS_wait	HTRANS must be stable after slave inserts a wait state.
err_M_HTRANS_idle_next	Transfer of type IDLE cannot be followed by type SEQ or BUSY.

7.3.5 Formalize Requirements Checklist

For the fifth step, using the interface signals identified in step 3 and the set of requirements identified in step 4, create your formal

TABLE 7.1

Interface Description

Signal Name	Description	Size	Master/Slave
HRESETn	Reset (active low)	1 bit	M/S
HCLK	Clock	1 bit	M/S
HREADY	Ready on AHB bus	1 bit	M/S
HRESP	Transfer response	2 bits	M/S
HADDR	Address bus	32 bits	M/S
HTRANS	Transfer type	2 bits	M/S
HWRITE	Transfer direction	1 bit	M/S
HSIZE	Transfer size	3 bits	M/S
HBURST	Transfer burst type	3 bits	M/S
HWDATA	Data bus for write	`HSIZE_BITS	M/S
HRDATA	Data bus for read	`HSIZE_BITS	M
HGRANT	Grant	1 bit	M
HBUSREQ	Request	1 bit	M
HLOCK	Master lock to arbiter	1 bit	M
HPROT	Protection control	4 bits	M
HREADYOUT	Ready from slave	1 bits	S
HSEL	Slave selected	1 bit	S
HMASTER	Granted master	4 bits	S
HMASTLOCK	Arbiter lock to slave	1 bit	S
HSPLITx	Transfer split	`MASTER_MAX	S

specification. The following PSL code formalizes the natural language requirements that were created for the requirements checklist. These requirements are easily specified through the use of a *forbidden* sequence.

```
//  HTRANS set to IDLE after reset
err_M_HTRANS_rst_value: assert never
    {(~HRESETn); ~(HTRANS==`IDLE)};

//  Transfer of type IDLE cannot be followed by types SEQ
//  or BUSY
//  @see  Table 3-1, p.3-9
err_M_HTRANS_idle_next: assert never
    {(HTRANS==`IDLE); ((HTRANS==`SEQ) || (HTRANS==`BUSY))};

//  Transfer of type BUSY cannot be followed by type NSEQ
//  when not INCR burst
//  @see p.3-9
```

```
err_M_HTRANS_busy_next: assert never
    {((HTRANS==`BUSY) && (HBURST!=`INCR)); (HTRANS==`NSEQ)};
```

7.3.6 Document Verification Strategy

In this step, you will define the verification plan and proof strategy. It has been our experience that it is usually easiest to start your AMBA AHB proof with the simpler rules. Choose a requirement that is easy to understand, has little or no interaction with other requirements, has few steps to set up, and is easy to analyze.

Note that this section would normally document advance techniques that would be employed to complete the proof. Chapter 8, "Techniques for Proving Properties," introduces a few advanced techniques.

7.3.7 Define Coverage Goals

In the final step, we define a set of coverage goals that you can use to confirm that your coverage objectives were met during the proof. Many formal tools allow you to verify that a boolean expression is valid on the analysis region after a proof. Hence, you can check to see if a particular state was ever encountered during the search process. This is useful for determining if the design was over constrained during the proof process, or if formal specification was incorrect (i.e., it prevented us from exploring a state we were expecting).

7.4 SUMMARY

This chapter introduced a set of steps necessary to create a formal test plan. It was our goal to provide sufficient knowledge that you can create customized formal test plan chapters for your own design. In addition, we introduced a set of guidelines for effective assertion-based specification techniques.

CHAPTER 8

Techniques for Proving Properties

In the ideal situation, you are delighted to find that your formal verification tool is able to automatically prove the specified properties. However, often you will be less than delighted and you will be wondering, "What should I do next when the proof is too complex for my formal tool?"

In this chapter, we introduce various techniques that address complexity. Whether you plan to verify a complex system by using simulation or formal verification, success in dealing with complexity often depends on a few fundamental concepts and advanced techniques, such as *abstraction*, *decomposition*, and *symmetry*. Some of the state reduction techniques we discuss can be automated and occur automatically under the hood of many formal engines. Other techniques are not automatic and require you to manually apply them. Obviously, your decision to apply these techniques depends on the importance of the properties in question and the expected return on investment for proving them.

8.1 CONE-OF-INFLUENCE REDUCTION

Cone of influence is an effective state reduction technique that is commonly employed by all commercial formal tools. The concept is illustrated in Fig. 8.1.

FIGURE 8.1

Cone of Influence

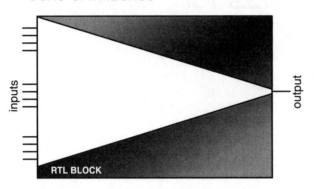

In Fig. 8.1, the state variables in the white region are contained in the cone of influence for a property specified on the output of the block. The state variables contained within the dark region do not contribute to the behavior of the RTL model with respect to the specified property. Hence, they can be removed from the analysis to reduce the state space required for the proof.

Modern formal verification tools automatically identify the set of state variables that potentially affect the property as a first step, and then they remove all the other variables from the analysis. Computing the variables contained in the *cone of influence* is straightforward. The tool begins by including the variables that directly occur in the property specification into the *cone-of-influence set*. Then for each variable v_i contained in the cone-of-influence set, the tool will identify all variables that appear on the right-hand side of the RTL assignments to this variable v_i, and add the new variables to the *cone-of-influence* variable set. This process is repeated until no new variables can be added to the set.

How to Apply This Technique
This is an automated technique performed by most formal verification tools today. Hence, there is nothing you need to do or set up to take advantage of this state reduction technique.

8.2 ABSTRACTION REDUCTION

Dealing with complexity by layering design representations is prevalent throughout all disciplines of engineering. Examples increasing abstraction in electronic design include the transistor model, gate model, RTL model, and behavior or architectural model. The common characteristic for all these examples is that higher levels of representation make assumptions or simplifications about a lower-level representation in order to make analysis tractable (e.g., a gate-level model is a simplification of a transistor model). This layering or simplification concept is known as *abstraction*. See Fig. 8.2.

In general, abstraction is the process of extracting the underlying essence of a concept, removing any dependence on real-world objects with which it might originally have been connected and generalizing it so that it has wider applications. For example, in electronic design, it is not uncommon to perform RTL simulation by using a simple unit delay model to represent time. This form of abstraction allows the engineer to focus on functional aspects of the design while ignoring the detailed timing information (which can be verified more efficiently in a separate process).

In formal verification, abstraction is a common technique used to address complexity. Many engineers fear the concept of abstraction. That is, they often fear that the abstraction might dangerously introduce an error that would lead to the property's proving true when it is really false (also referred to as a *false positive*). However, there is a

FIGURE 8.2

Layers of
Representation

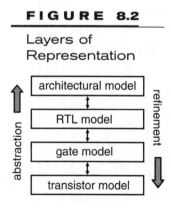

class of abstractions known as *safe abstractions*, where if the property is proved true on the abstracted model, then the property is guaranteed to be true on the real model. The reason is that (for this class of safe abstractions) the behavior described by the abstract model is a superset of the actual behavior described by the real model.

For example, consider the design illustrated in Fig. 8.3. In this design, A and B are always inverted (that is, {A,B} = 01 or {A,B} = 10). The actual values of A and B are contained within the design's reachable state space, as indicated by the dark shape in Fig. 8.4.

At times, it is actually easier for a formal tool to prove a property on a superset of the reachable state space, as indicated in the inner rectangle in Fig. 8.4.

FIGURE 8.3

Reachable States
of Signals A and B

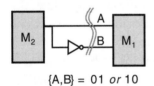

{A,B} = 01 *or* 10

FIGURE 8.4

Superset of the Reachable State Space for AB

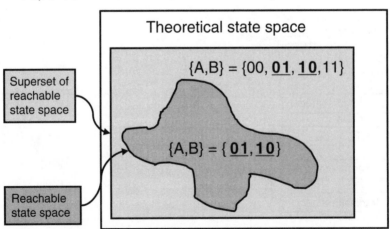

To illustrate this concept, consider the example shown in Fig. 8.3. If we make a cut at signals A and B in the circuit, and we then treat AB as inputs to the design, allowing them to assume during a proof any of the values

$$00, 01, 10, 11$$

then the simpler circuit is actually an abstraction of the larger circuit. Note that the behavior of the larger circuit (where AB can only assume the values $\{A,B\} = 01$ or $\{A,B\} = 10$) is contained within the behavior of the simpler circuit created by the cut (where $\{A,B\} = 00$ or $\{A,B\} = 01$ or $\{A,B\} = 10$ or $\{A,B\} = 11$). This concept is illustrated in Fig. 8.4 as the superset of the reachable state space.

If we prove that a safety property is true on the simple circuit M_1 (with the cut at A and B), then it is not necessary to prove the property on the larger circuit that contains both M_1 and M_2 (since, as we previously indicated, the behavior of the larger circuit is contained within the behavior of the simpler circuit with respect to A and B). Hence, the smaller circuit used during the analysis simplifies the proof. However, if the property is not proved true on the simple circuit, then we cannot conclude that M_1 is in error. Under this circumstance, it becomes necessary to reprove the property on the larger circuit without the cut.

How to Apply This Technique

Some formal verification tools automatically apply this technique of abstraction. That is, many tools will heuristically identify a cut in the design (as shown in Fig. 8.3), and attempt to prove the design on the abstract model. If these prove false, these tools will attempt to refine the abstract model (i.e., include additional logic driving the boundary of the cut to eliminate the false negative).

You can manually apply this technique yourself for complex blocks on a coarser-grain cut (i.e., hierarchical boundaries). For example, if you have a large block containing multiple smaller blocks, you might attempt to prove a property on the smaller block, without including the complete logic of the larger block. This is a form of compositional reasoning, which is discussed next.

8.3 COMPOSITIONAL REASONING

Compositional reasoning is a process that reduces reasoning about a larger concurrent system to reasoning about its individual components. This technique is essential for managing proof complexity and state explosion in model checking. Note that compositional reasoning transfers the burden of proof from the global level to the local component level so that global properties can be inferred from independently verified component properties. In this section, we will explore various forms of compositional reasoning.

8.3.1 Unconstrained Decomposition

For designs with very large state spaces, prior to the application of a formal verification, often the designs need to be decomposed into subcomponents to manage complexity. In fact, identifying the decomposition is often straightforward by simply following the component structure of the design. Using this decomposition, we can at times prove properties on the smaller pieces while abstracting out other components. To illustrate this concept, consider the design illustrated in Fig. 8.5.

To simplify our proof, we could remove the Parity Check subblock, which creates an abstraction of the original circuit by allowing `par_err` to assume any value at any time. Hence, while proving property p, we will explore all possible combinations of `par_err` with respect to the input data bus. When we have complete the proof

FIGURE 8.5

Decomposition

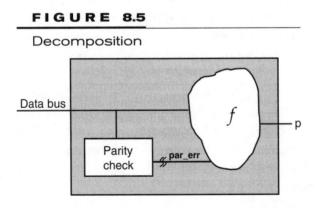

of p, we then prove the parity circuit functions correctly as a separate proof.

How to Apply This Technique

You can manually apply this technique yourself on blocks that contain complex functional components, such as parity checking. Create a cut on the output of the complex function, and allow the cut to assume all possible sequences of values during your proof.

8.3.2 Assume-Guarantee Reasoning

Assume-guarantee reasoning is a form of compositional reasoning in which a property is proved on a decomposed component using a set of assumptions about other neighboring components, and then these assumptions are proved separately on the neighboring components, as illustrated in Fig. 8.6.

In this example, we first prove that property P holds on block C, using assertions A_1 and A_2. Hence, it is unnecessary to include block A and block B as part of the proof, which reduces the state space in the analysis region. We then prove that assumption A_1 holds on block A, and we prove that assumption A_2 holds on block B.

How to Apply This Technique

You can manually apply this technique yourself on complex blocks with clearly defined interface. Figure 8.6 illustrated the concept of assume-guarantee reasoning, where assertions A_1 and A_2 are

FIGURE 8.6

Prove Property P using assumptions A_1 and A_2

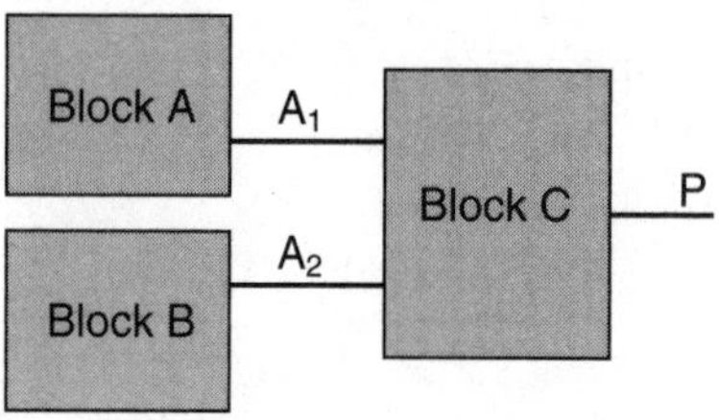

assumptions to block C, and then must be proved on block A and block B.

8.4 SYMMETRY

Many types of designs (such as pointer-based, round-robin arbiters) have a symmetric implementation. That is, the logic associated with each port has the exact same RTL structure. Generally, the only exception to a symmetric arbiter implementation is the initial priority value for each port (i.e., you will give one port the highest priority).

Symmetry allows us to efficiently prove properties on a pair of symmetric paths (e.g., ports of an arbiter) while giving us confidence that the requirement is true on all ports due to the symmetric implementation. If you are confident about the symmetry among ports in the implementation of your design, then we recommend you take advantage of this symmetry as part of your proof.

Consider the following Verilog code fragment for an arbiter that has a symmetric implementation per port with the exception of the initial priority value:

```
// pointer to determine round-robin
reg [3:0] priority_pointer; // port
reg [7:0] grant;
always @(posedge clk) begin
  if (reset)
  begin
    priority_pointer <= 4'd0;
    grant <= 8'd0;
  end
  else if (grant !=8'd0) grant <= 8'd0;
  else if (request[priority_pointer])
  begin
      priority_pointer <=
                priority_pointer+5'd1;
    grant[priority_pointer] <= 1'b1;
  end
  else if (request[priority_pointer+5'd1])
  begin
    priority_pointer <=
```

```
            priority_pointer + 5'd2;
    grant[priority_pointer + 5'd1] <= 1'b1;
  end
  else if (request[priority_pointer+5'd2])
  begin
    priority_pointer <=
                priority_pointer + 5'd3;
    grant[priority_pointer + 5'd2] <= 1'b1;
  end
  .  .  .  .  .
  end
```

In the previous Verilog example, you can see that the logic associated with each port is symmetric since each n-bit variable assignment follows the same RTL structural path. The only part of this implementation that is not symmetric is the reset value of priority_pointer, which points to port0 after reset (i.e., it has the highest priority after reset).

One technique that can be used to prove this circuit by taking advantage of symmetry is to ignore the initial value of priority_pointer (either by modifying the circuit or taking advantage of some options available in certain formal tools). This technique permits all possible initialization values for the signal priority_pointer to be explored, which restores the true symmetry to the circuit. Hence, we can prove properties, such as fairness, on the circuit by only having to examine an arbitrary pair of grants (since each grant is symmetric). This greatly simplifies the proof.

Other examples of symmetry include data paths within a block, which generally have implied a symmetric property on the data bus. For example, consider an 8-bit data transport bus where the entire bus is referenced within the DUT without individually routing bits. In addition, no indiviual bits within the data bus are uniquely assigned. Hence, proving a property (such as packet data will not be dropped, duplicated, or corrupted) on a single bit of the bus is sufficient to guarantee that the entire bus is correct.

How to Apply This Technique
If you are attempting to prove a data integrity property (e.g., data on an n-bit bus are not corrupted, dropped, or duplicated from the

input to the output of a block), then a large state reduction will occur if you specify your property with respect to a single bit of data (versus all bits of data). In many cases, you can take advantage of the symmetry of the data path. When you are not sure if the data path is symmetric, you should still partition the property into separate properties for each bit (versus attempting to specify a single property that describes all n bits).

8.5 COUNTER ABSTRACTION

Counters are one of the most common components found in a design. If a counter is large, it can be a challenge for any formal verification tool since it forces the number of evaluation iterations to be very high, without necessarily exploring new interesting states.

For a design involving an arbiter, there are several places in the arbiter implementation that often require counters. Dynamic arbiters, e.g., typically use a counter to keep track of the arbitration change events. Fortunately, it is usually possible to abstract these counters to reduce the reachable states without compromising the proof.

There are two main counter abstractions: *counter induction* and *counter reduction*.

8.5.1 Counter Induction

The first abstraction we discuss uses induction to abstract out the details of the counter. For example, consider a dynamic priority arbiter that requires us to prove that the credit/weight is incremented, reset, and decremented correctly. If we follow the actual logic where the credit is reset to a fixed value, the formal engine will need to go through many cycles to explore all possible credit values. Alternatively, we can partition the requirement into two parts:

1. The credit register is initialized correctly.
2. The credit register is incremented and decremented correctly.

The first requirement can be specified as follows:

```
// credit_init is the expected
// initialization value of credit and
// prev_rst is the registered version of
// reset, hence indicate the value of
// credit during the very first cycle
assert never
    (prev_rst & (credit!=credit_init));
```

The second set of requirements can be specified as follows:

```
// error credit increment
assert never
  (credit_inc & (credit!=prev_credit+1));

// error credit decrement
assert never
  (credit_dec & (credit!=prev_credit-1));
```

In addition, for the second set of requirements you must let the initial credit value be free (i.e., it is uninitialized and allowed to take on any value). Hence, by freeing the credit values for the second set of requirements, we are very efficiently checking that the increment/ decrement is working correctly for all possible values of credit, without having to sequentially iterate through the credit count sequence.

8.5.2 Counter Reduction

Use the second type of counter abstraction when a specific counter value triggers a specific event (e.g., a time-out counter). Since most of the counter values are uninteresting and are used only as a sequence to generate the next interesting value, we can perform an abstraction to reduce the number of states that the formal engine needs to evaluate.

For example, assume our design has a counter that calculates the elapsed time before the design must refill its credit for a credit-based arbiter. For this type of design, there are only two counter values that are interesting (that is, 0 and the time-out count to generate the credit refill). All other counter values have no influence or impact on the rest of the design. Hence, we can use an abstraction technique that will fast-forward the counter values that are uninteresting and thus reduce the state space needed to prove the design's high-level requirements.

Original counter model:

```
reg [15:0] count;
always @(posedge clk) begin
  if (rst) count <= 16'h0;
  else if (count == credit_refresh_cnt)
          count <= 16'h0;
  else    count <= count + 16'h1;
end
```

Abstracted counter model:

```
reg         abstract_cnt;
reg [15:0] nxt_count;
always @(posedge clk) begin
  if (rst) abstract_cnt <= 1'b0;
  else if (count==credit_refresh_cnt)
          abstract_cnt <= 1'b0;
  else    abstract_cnt <= 1'b1;
end
```

For the abstracted counter, add the following to assumptions at a cut point in your original counter:

```
// count can assume any value less than or
// equal to credit_refresh_cnt
assume always (abstract_cnt ->
          (count<=credit_refresh_cnt));

// count will be set back to zero
assume always
      (~abstract_cnt -> (count == 16'h0));
```

In our previous example, during reset, the value of abstract_cnt is set to 0, which in turn sets the count to 0 with the specified assumption. After that, abstract_cnt is set to 1, which allows the count to assume any (and all) values between 0 and the credit_refresh_cnt with the specified assumption. If there are no meaningful states for the formal tool to analyze between 0 and `credit_refresh_cnt`, then the count will not arbitrarily assume a value that is less than `credit_refresh_cnt` for any time longer than necessary. Hence, count will assume `credit_refresh_cnt` (an interesting state) sooner without having to sequence through all the counter values. When count assumes the value of `credit_refresh_cnt`, `abstract_cnt` resets to 0, which in turn resets count to 0.

One advantage of this type of counter abstraction is that if we miss some meaningful counts (counts that impact the rest of the design) when we are creating our abstract model, our proof returns a *false* with a counter example. Thus, this form of abstraction might give you a false negative, but it does not give you a false positive. If there are more counter values needed in our abstracted model, then we can declare more states and map these states onto states in our abstracted model. This type of counter abstraction is also useful when counting the number of remaining credits in the credit-based dynamic arbiter.

How to Apply This Technique

If you are unsuccessful at proving a property on your block, examine the block to determine if there is a counter in the cone-of-influence analysis regions that could be abstracted. For some counters, you can simply introduce a cut in the design and prove the design for any value of the counter (while proving that the counter counts correctly as a separate proof). For other counters, where specific count values trigger certain events, we suggest you try the counter reduction technique previously discussed.

8.6 NONDETERMINISM

Specifying behavior in formal verification often is quite different from techniques for monitoring behavior in a simulation testbench. For example, in simulation, often counters are introduced to count cycles between events. This technique can present problems for formal tools. A better approach in formal tools is to use nondeterminism in specifying the relationship of events.

For example, consider the sequencing through states A, B, C, and D, where in simulation a count-down counter is used to determine the transition to the next state. This could be modeled in formal verification as follows:

```
reg [3:0] state
always (posedge clk) begin
  if (rst_n==0) state <='A;
  else state <= n_state;
end

assume ((state=='A) ->
          (n_state=='A)||(n_state=='B));
assume ((state=='B) ->
          (n_state=='B)||(n_state=='C));
assume ((state=='C) ->
          (n_state=='C)||(n_state=='D));
assume ((state=='D) ->
          (n_state=='D)||(n_state=='A));
```

Note that the state transitions are actually modeled through a set of assumptions. Also note the use of nondeterminism in the model. For example, if the FSM state is in the 'A state, then the assumptions allow it to either stay in that state or nondeterministically change to state 'B. Hence, it is unnecessary to introduce a counter in this model. The formal engine will explore a range of all possible temporal relationships of the state 'A to 'B transitions, efficiently using nondeterminism.

How to Apply This Technique
Look for instances in your specification where you are attempting to explicitly count events, and determine whether you can implicitly model the behavior using nondeterminism.

8.7 GRADUAL EXHAUSTIVE FORMAL VERIFICATION

In some cases, engineers prefer to verify various functionality or modes of operation for their design separately. This might be due to the engineers' desire to start an early verification on an incomplete RTL model where some functionality is complete while other functionality remains partially coded. Another reason to perform verification on separate design modes stems from a sense of familiarity. That is, for traditional simulation-based methodologies, the engineer might partition the development of a testbench into separate stimulus generators for various design operating modes. For example, for a design containing a USB interface, individually a host and a device must be able to handle both normal and error conditions. If the engineer starts the verification with both normal and error conditions, then it is likely that too many bugs will be detected for the error condition. This can frustrate the designer who would not have a sense of whether the basic functionality for the normal condition was working correctly. Hence, the engineer might take the following course of action in a traditional simulation-based methodology to allow partitioning of the various operating modes during verification:

1. Develop a generator for normal condition transactions.

2. Begin verification for this single mode of operation.

3. When the testbench is no longer detecting mainstream bugs associated with normal condition transactions, develop a generator for error condition transactions.

4. Perform verification on this mode of operation.

5. After sufficient verification has occurred on error condition transactions, perform the verification by combining random occurrences of normal and error condition transactions within the testbench with the goal of flushing out corner-case bugs.

You can apply a similar methodological approach when using a formal verification tool such as a model checker. This approach is referred to as *gradual exhaustive formal verification via restrictions*, and it has the potential of flushing out mainstream bugs as quickly as possible while using formal verification to search a large state space.

This approach offers the following benefits:

♦ It enables you to initially turn off portions of the design's functionality—and then gradually turn on additional functionality as you validate the design under a set of restrictions. (*Note*: This is analogous to creating separate testbench generators for simulation-based verification.)

♦ It provides an easier method of debugging by selecting, and thus controlling, the functionality in the environment that is enabled.

♦ It allows you to refine the constraint model (i.e., assumptions) into more general assumptions without initially encountering state explosion.

Essentially, this approach involves gradually developing a formal verification environment around the RTL component by using restrictions. A *restriction* is a special type of constraint, in that it constrains the design behavior explored by the formal verification tool to a given artificial assumption. For example, a restriction may reduce a large set of opcodes to a smaller set of opcodes to be explored during the formal search process. Or, as with a traditional simulation approach, a restriction may limit the input behavior to only *read* transactions during one phase of a proof, and then reprove the design for *write* transactions. Other examples include restricting the upper 8 bits of a 16-bit bus to a constant value while letting the lower 8 bits remain unconstrained during the formal search, then shifting the restriction to a new set of bits and reproving with the new bus restrictions.

Figure 8.7 illustrates the restricted state space concept. The outer square represents the entire or theoretical state space, which consists of the reachable as well as unreachable state space. The entire state space consists of the following number of theoretical states:

$$(R + I)^2$$

where R is the number of state elements found in the design and I is the number of input signals into the design. In general, not all possible combinations of input values are possible or legal. Similarly, not every possible combination of state element values is possible. Hence, the dark region represents the reachable (i.e., legal) state space associated with the design. Note that if we further restrict the input value (e.g., limit input values to only read-type transactions), then the behavior of the design considered during

FIGURE 8.7

Restricted State Space

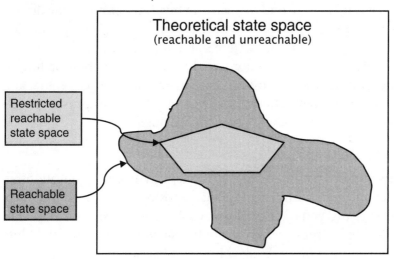

verification is simpler and reduced, as illustrated by the light gray region in Fig. 8.7. The technique we discuss in this application note uses restrictions to reduce the behavior analyzed during formal verification with the goal of targeting simpler mainstream bugs early in the design process.

One characteristic of a restriction is that it cannot be proved on a neighboring block (that is why we referred to the restriction as an artificial assumption). For example, if we restrict all input sequences to only read transactions, proving this restriction on a neighboring block would fail since write transactions could also be generated (since the neighboring block could produce both read and write transactions). Nonetheless, restrictions enable the engineer to narrow the verification process and quickly find many mainstream bugs. Even with the use of restrictions, the number of scenarios that the formal verification tool explores is very large; however, the formal verification tool will detect complex errors under these conditions as well.

After we have formally verified the design using various restrictions, our next step is to relax the restrictions into more general interface assumptions. By doing this, we are able to prove the general assumptions as properties on neighboring blocks. Note the subtle distinction between assumptions and restrictions. For restrictions, our goal is to quickly find mainstream bugs and clean up the design by using formal techniques. We are under no obligation to validate restrictions (neither in simulation nor in formal verification). In fact, it is likely that the restriction would fail in simulation or formal verification since it is an overconstraint and too restrictive. In using assumptions as restrictions, however, our goal is to prove correctness while finding complex corner-case bugs.

AMBA AHB Example

In this section, we illustrate this concept by using assumptions as restrictions on a design containing an ARM AMBA AHB interface. To restrict the behavior explored by the formal tool for the AHB interface to a simple, single write transaction consisting of 32 bits with no address offset, we key in the PSL assumption:

```
assume always (HSELS -> ((HADDRS[3:0] ==
4'b000) & \
              (HBURSTS == 3'b000) & (HSIZES ==
3'b010) & \
              (HWRITES == 1'b1) &
(HREADYM==1'b1));
```

The various subexpressions describe the following behavior:

```
HADDRS[3:0] == 4'b000 # no address offset
HBURSTS == 3'b000 # single transaction
HSIZES == 3'b010 # 32-bit transfer
HWRITES == 1'b1 # write transaction
HREADYM==1'b1 # output always ready
```

For clarity, this complex assume command could be coded as a set of simpler assume commands, where each command represents a unique restriction as shown:

```
AHB_NO_ADDR_OFFSET: assume always
                (HSELS -> HADDRS[3:0] == 4'b000);
AHB_SINGLE_TRANS: assume always
                (HSELS -> HBURSTS == 3'b000);
      .   .   .
```

After we successfully prove properties on our design by using the previous restrictions, the idea is to gradually remove (or relax) the various restrictions (artificial assumptions), until we are proving properties on our design with only general real assumptions.

The restriction for a single transaction could now be relaxed to handle other modes, such as another INCR burst length, as demonstrated below:

```
assume always
    ((HBURSTS==3'b000) | (HBURSTS==3'b101) | \
    (HBURSTS==3'b011) | (HBURSTS==3'b111));
```

After reproving our properties with the relaxed restriction, we could remove the restriction that the output be always ready (HREADYM == 1'b1) and then reprove our set of properties. We then might choose to remove the write transaction restriction (RWRITES == 1'b1) and reprove our set of properties. Again, our goal is to gradually remove all restrictions until we are left with real input assumptions.

8.8 SUMMARY

In this chapter, we introduced few fundamental concepts and advanced techniques that can be effectively used to address complexity, such as *abstraction, decomposition,* and *symmetry.* Some of the state reduction techniques we discussed can be automated and occur automatically under the hood of many formal engines. Other techniques we discussed are not automatic and require you to manually apply them. Obviously, your decision to apply these techniques depends on the importance of the properties in question and the expected return on investment for proving them.

CHAPTER 9

Final System Simulation

Up to this point we have talked about how important it is to verify devices and the techniques used to perform the verification. We have concentrated most of the effort on formal techniques, as that is the topic of this book.

9.1   TEST PLAN REVISITED

We have examined some of the ways that the test plan is used to form the overall strategy for verifying the device. For large designs the design was split into multiple modules, and the modules were verified separately. The test plan is instrumental in specifying exactly how each module will be verified. Each module of the design has a targeted approach to its verification. This includes the types of tests that need to be run against the module, the expected results, and the tool used to do the verification. The test plan will guide the efforts of the design and verification engineers so that they can create realistic estimates for how much effort will be involved to verify a module and the steps necessary to perform the verification.

9.2 MODULE VERIFICATION

Most of the verification effort up to this point has focused on making sure that a particular module of the design behaves as expected. The designer most likely has used a number of different tools to verify different modules. Depending on the tool used, the design and verification engineer has spent a lot of time creating simulation testbenches or formal verification assumptions and requirements. These testbenches and assumptions have been targeted at the particular module in the design being verified.

9.3 FULL SIMULATION FROM A SIMULATION FLOW

Simulation was used initially by the designer to verify basic module functionality. Comprehensive testbenches were then created by the design and verification engineer to verify that a particular module was correct. These testbenches were run against the module, and the module functionality was modified until it behaved correctly per the testbenches.

Figure 9.1 is a block diagram of a design with four modules. Designers using a simulation-based approach to verification would

FIGURE 9.1

Full Design

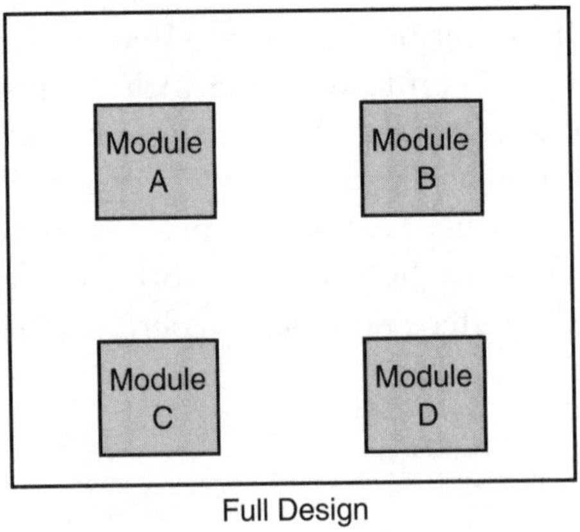

Full Design

FIGURE 9.2

Module-Level Verification

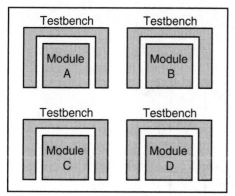

Full Design

create testbenches for each of the modules and verify module by module. When all modules were verified, the design would look as shown in Fig. 9.2.

Each module has a testbench or set of testbenches that verifies the functionality of the module. At this point the designer is confident that the module is correct, but the overall design could still have errors. These errors could be module interconnection errors or module interaction errors. The designer needs to perform a simulation of the complete design to verify that none of these errors exist.

The designer will need to create a full system simulation that includes all the modules of the design and a full system simulation testbench. This is shown in Fig. 9.3.

This setup will verify that the full system is behaving as expected. However, the designer needs to ensure in the full system simulation testbench that the functionality verified during module verification is also verified. The designer will need to create the full system testbench such that it incorporates the module test functionality. This is shown in Fig. 9.4.

This is not an automatic process unfortunately, so it will require more work by the verification team to make this happen.

FIGURE 9.3

Full System-Level Verification

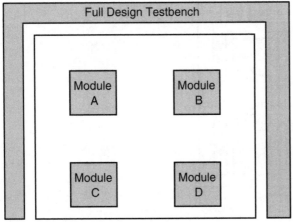

Full Design

FIGURE 9.4

Module-Level and Full System-Level
Verification

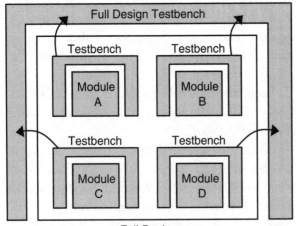

Full Design

9.4 FULL SIMULATION FROM A FORMAL VERIFICATION FLOW

If the verification team decided to use formal verification to test the modules, then an alternate strategy must be used to perform the full system simulation. If the module contained complex control functionality, the designer might have used formal verification to verify the modules. Formal verification produces modules that are very free of errors. Modules that were formally verified would probably not have comprehensive testbenches, but instead would have formal requirements and assumptions.

Not all blocks would be formally verified. As discussed earlier in the book, some modules are not easily verified with formal verification. The design would contain a mix of modules, some that were formally verified and others that were simulated. This is shown in Fig. 9.5.

Modules A, B, and C have been formally verified while module D has been simulated. Module D has a testbench associated with it, while modules A, B, and C have assumptions that need to be verified in the full system environment. Assumptions between formally

FIGURE 9.5

Mixed Verification Methodology

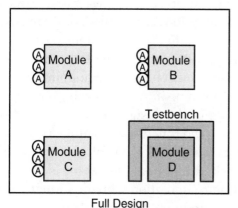

Full Design

☐ Formally Verified   Ⓐ Formal Assumptions

FIGURE 9.6

Assume-Guarantee Methodology

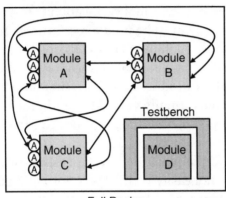

Full Design

☐ Formally Verified   Ⓐ Formal Assumptions

verified modules have been verified with an assume-guarantee methodology, as discussed earlier. This is shown by Fig. 9.6.

However, each module assumption still needs to be verified in the total system environment. The designer will need to create a full system testbench that includes the functionality of the module testbenches but also verifies the module assumptions used for formal verification. This is shown in Fig. 9.7.

To make the full system verification process easier and to determine how well the module assumptions are verified, designers will insert functional coverage points into the full system simulation. *Functional coverage points* are monitors that listen to points in the design. The designer will insert functional coverage points at specific points in the design to determine if the testbench exercises all the desired functional behavior of the design. If the desired functional behavior is not exercised, the functional coverage point will report that fact to the verification engineer, who will need to modify the testbench so that the desired functional behavior is verified. Figure 9.8 shows the final system simulation with the formally verified modules, the simulated modules, the formal assumptions, and the functional coverage points inserted.

FIGURE 9.7

Formal Assumptions

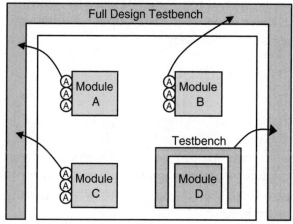

Full Design

☐ Formally Verified   Ⓐ Formal Assumptions

FIGURE 9.8

Functional Coverage

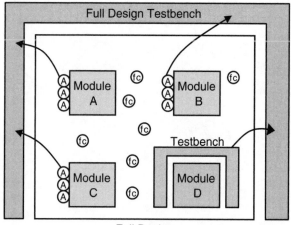

Full Design

☐ Formally Verified   Ⓐ Formal Assumptions   ⓕⓒ Functional Coverage
                                                      Point

The verification team will run the final system simulation with the full design testbench and use the results to determine whether the design has the correct functionality. Once the design is functionally correct, it can be released to the group who will implement the design in the target technology.

9.5 SUMMARY

It is very important that a consistent methodology be followed to verify a design. This also means using the best tools at hand for the task. In the module verification process, formal verification can produce modules that have few, if any, errors. This can have a big impact on the time for final system simulation, where finding errors in modules is much harder.

IEEE 1850 PSL Property Specification Language

A.1 INTRODUCTION TO IEEE 1850 PSL

In this appendix, we discuss the proposed IEEE 1850 PSL property specification language, which is a formal property language donated to IEEE by Accellera. This language was based on the Sugar property language originally developed by IBM and is compatible with multiple HDLs (i.e., the user can specify properties to work across Verilog, SystemVerilog, and VHDL). PSL was developed to address these shortcomings of natural language forms of specification. It gives the design architect a standard means of specifying design properties using a concise syntax with clearly defined formal semantics. Similarly, it enables the RTL implementer to capture design intent in a verifiable form, while enabling the verification engineer to validate that the implementation satisfies its specification through dynamic (i.e., simulation) and static (i.e., formal) verification means. Furthermore, it provides a means to measure the quality of the verification process through the creation of functional coverage models built on formally specified properties. Plus, it provides a standard means for hardware designers and verification engineers to rigorously document the design specification (machine-executable).

The PSL property language is suited for specifying architectural properties prior to *and* during RTL implementation. In addition, as a declarative form of specification, PSL is suited for specifying interface properties during block-level refinement from an architectural model. Finally, the expressiveness of PSL makes it excellent for capturing implementation assertions and boundary assumptions during RTL implementation.

PSL is a comprehensive language that includes both linear temporal logic (LTL) and computation tree logic (CTL) constructs. This enables PSL to support various kinds of formal verification flows, including event-driven and cycle-driven simulation and various algorithms for formal verification (model checking). However, while every PSL property can be checked in formal verification, not every PSL property can be checked in simulation. A subset of the PSL language lends itself to automatically generating simulation checkers (also referred to as *monitors*), which then can be used to check specific design properties during simulation. Hence, we are limiting our PSL introduction to a subset of the language that can be checked in both simulation *and* formal verification. A full description of the PSL language can be found in Accellera proposed standard Property Specification Language (PSL) 1.1 [Accellera PSL-1.1 2004].

A.2 OPERATORS AND KEYWORDS

Table A.1 shows PSL keywords, which are case-sensitive. This appendix discusses the keywords marked in **bold** type.

Table A.2 lists the operators available in PSL from highest to lowest precedence. This appendix addresses the operators in bold. Those in regular type are beyond the scope of this appendix. Additional information on these operators can be found in the PSL Language Reference Manual (LRM) [Accellera PSL-1.1 2004].

PSL also defines a set of *options branching extension* (OBE) operators, which are beyond the scope of this appendix. Please refer to the PSL LRM [Accellera PSL-1.1 2004] for details.

TABLE A.1

PSL Keywords

A	E	next_a!	to†
AF	EF	next_e	
AG	EG	next_e!	
AX	EX	next_event	U
abort	endpoint	next_event!	union
always	eventually!	next_event_a	until
and†		next_event_a!	until!
assert	F	next_event_e	until!_
assume	fairness	next_event_e!	until_
assume_guarantee	fell	not†	
	forall		vmode
before		onehot	vprop
before!	G	onehot0	vunit
before!_		or†	
before_	in		W
boolean	inf	property	within
	inherit	prev	
clock	is†		X
const	isunknown	report	X!
countones		rose	
cover	never		
	next	sequence	
default	next!	stable	
	next_a	strong	

† Keyword only in the VHDL flavor.

A.3 PSL BOOLEAN LAYER

The *boolean layer* of PSL provides for any boolean expression valid within the language flavor of PSL being used (i.e., Verilog, SystemVerilog, or VHDL boolean expressions). The result of the boolean expression is a singular value of *true* or *false*. This is equivalent to a condition being evaluated within an if statement within Verilog or VHDL. Additionally, PSL provides a number of predefined functions that return boolean values.

TABLE A.2

PSL Operator Precedence

Operator Class	Associativity	Operators
(*Highest precedence*) HDL operators		
Union	Left	union
Clocking	Left	@
SERE repetition	Left	[*] [+] [=] [->]
Sequence AND	Left	& &&
Sequence OR	Left	|
Sequence fusion	Left	:
Sequence concatenation	Left	;
FL termination	Left	abort
FL occurrence	Right	next* eventually!
		X X! F
FL bounding	Right	U W
		until* before*
Sequence implication	Right	|-> |=>
Boolean implication	Right	-> <->
FL invariance	Right	always never
(*Lowest precedence*)		G

A.4 PSL TEMPORAL LAYER

The *temporal layer* is the heart of the PSL language; it describes properties of the design that have complex temporal relationships. Thus, unlike simple properties such as `signals a and b are mutually exclusive`, the temporal layer allows PSL to describe relationships between signals, such as `if signal c is asserted, then signal d must be asserted before signal e is asserted, but no more than 8 clock cycles later`.

PSL's *temporal layer* is based on an extension of regular expressions, called *sequential extended regular expressions* (SEREs), which is in many cases more concise and easier to read and write. The simplest SERE is a boolean expression describing a boolean event. More complicated SEREs are built from boolean expressions using various SERE composition operators. A SERE is not a property on its own; it is a building block of a property. That is, properties are built from temporal operators applied to SEREs and boolean expressions.

Within this section, we refer to a sequence *holding*. This term indicates that the behavior described by the sequence or property is actually seen.

This section is organized by describing the composition operators first, followed by the temporal operators.

A.4.1 SERE

Sequential extended regular expressions (SEREs), shown in Example A.1, describe single- or multicycle behavior built from a series of boolean expressions.

The most basic SERE is a boolean expression.

E X A M P L E A.1

Sequential Extended Regular Expression (SERE)

```
SERE  ::=
          Boolean
        | Sequence
        | Sequence_Instance
        | SERE ; SERE
        | SERE : SERE
        | Compound_SERE

Compound_SERE  ::=
     Repeated_SERE
   | Braced_SERE
   | Clocked_SERE
   | Compound_SERE | Compound_SERE
   | Compound_SERE & Compound_SERE
   | Compound_SERE && Compound_SERE
   | Compound_SERE within Compound_SERE
```

More complex sequential expressions are built from boolean expressions using various SERE operators. These operators are described in the subsections that follow.

A.4.2 Sequence

A *sequence* is a SERE that can appear at the top level of a declaration, directive, or property (see Example A.2).

A.4.3 Braced SERE

A SERE enclosed in braces is another form of sequence (see Example A.3).

A.4.4 SERE Concatenation (;)

The SERE *concatenation* operator ; describes a sequence of events by specifying two sequences of events that must hold one after the

EXAMPLE A.2

Sequence

```
Sequence  ::=
            Sequence_Instance
            | Repeated_SERE
            | Braced_SERE
            | Clocked_SERE
```

EXAMPLE A.3

Sequence

```
Braced_SERE   ::=
    { SERE }
```

E X A M P L E A.4

Concatenation of Sequences

```
SERE ::=
         SERE ; SERE
```

other. That is, the second SERE starts 1 cycle after the first SERE completes.

The right operand of ; is a SERE that is required to hold after the left operand completes. If either operand describes the empty sequence, then the concatenation holds if and only if the nonempty sequence holds. See Example A.4.

A.4.5 Consecutive Repetition ([*])

The SERE *consecutive repetition* operator [*] describes repeated consecutive concatenation of the same SERE. See Example A.5. Note the RANGE_SYM is : for Verilog and SystemVerilog and to for VHDL.

The first operand of consecutive repetition is a SERE that is required to hold several consecutive times. The second operator is a number (or a range of numbers) that describes the number of times the SERE is repeated.

If the high value of the range is not specified (or is specified as inf), then the SERE must hold for at least the low value of the range. If the low value of the range is not specified (or is specified as 0), then the SERE must hold for no more than the high value of the range times. If neither of the range values is defined, then the SERE is allowed to hold any number of times including zero; i.e., the empty sequence is allowed.

When there is no SERE operand and only a number (or a range), then the resulting SERE describes any sequence whose length is described by the second operand as above.

EXAMPLE A.5

SERE Consecutive Repetition

```
Repeated_SERE ::=
          Boolean [* [ Count ] ]
        | Sequence [* [ Count ] ]
        | [* [ Count ] ]
        | Boolean [+]
        | Sequence [+]
        | [+]
        | Boolean [= Count ]
        | Boolean [-> [ positive_Count ] ]
Count ::=
          Number
        | Range

Range ::=
          Low_Bound RANGE_SYM High_Bound

Low_Bound ::=
          Number
        | MIN_VAL

High_Bound ::=
          Number
        | MAX_VAL
```

The notation + is a shortcut for a repetition of one or more times. See Example A.6.

A.4.6 Nonconsecutive Repetition ([=])

Nonconsecutive repetition allows for space between the repetition terms. The syntax for nonconsecutive repetition is the same as for consecutive repetition except the * operator is replaced with the = operator. See Example A.7.

Note the RANGE_SYM is : for Verilog and to for VHDL.

EXAMPLE A.6

SERE Consecutive Repetition

```
SERE_A [ Number_n]              // SERE_A repeats exactly Number_n times

SERE_A [Number_n : Number_m]    // SERE_A repeats at least Number_n times
                                // no more than Number_m times

SERE_A [0 : Number_m]           // SERE_A is either empty or repeats no
                                // more than Number_m times

SERE_A [Number_n : inf]         // SERE_A repeats at least Number_n
                                // times

SERE_A [0 : inf]                // SERE_A is either empty or repeats some
                                // undefined number of times

SERE_A [+]                      // SERE_A evaluates one or more times

[* Number_n]                    // the sequence is of length Number_n

[* Number_n : Number_m]         // the length of the sequence is a number
                                // between Number_n and Number_m

[* 0 : Number_m]                // an empty sequence or a sequence of
                                // length Number_m at most

[* Number_n : inf]              // a sequence is of length Number_n at
                                // least

[* 0 : inf]                     // any sequence of events

[*]                             // any sequence of events (including the
                                // empty sequence)

[+]                             // any sequence of events of length 1
                                // at least
```

A.4.7   Goto Repetition ([->])

Goto repetition allows for space between the repetition of the terms. The repetition ends on the boolean expression being found. This facilitates searching for a particular expression and then continuing the sequence at the point it is found (see Example A.8).

EXAMPLE A.7

SERE Nonconsecutive Repetition

```
SERE   ::=
          Boolean [ = Count ]

Count  ::=
          Number | Range
Range   ::=
          LowBound RANGE_SYM HighBound
LowBound   ::=
          Number | MIN_VAL
HighBound   ::=
          Number | MAX_VAL
```

EXAMPLE A.8

Goto Repetition of a Sequence

```
SERE   ::=
          Boolean [-> [ positive_Count ] ]

Count  ::=
          Number | Range
Range   ::=
          LowBound RANGE_SYM HighBound
LowBound   ::=
          Number | MIN_VAL
HighBound   ::=
          Number | MAX_VAL
```

Note the RANGE_SYM is : for Verilog and SystemVerilog and to for VHDL.

A.4.8 Sequence Fusion (:)

The *sequence fusion* operator specifies that two sequences overlap by 1 cycle. In this case, the second sequence starts the cycle that the first sequence ends (see Example A.9).

EXAMPLE A.9

Sequence Fusion Operator

```
SERE ::= SERE : SERE
```

EXAMPLE A.10

Sequence Non-Length-Matching And Operator

```
SERE ::= SERE & SERE
```

EXAMPLE A.11

Sequence Length-Matching And Operator

```
SERE ::= SERE && SERE
```

A.4.9   Sequence Non-Length-Matching And (&)

The *sequence non-length-matching and* operator specifies that two sequences must hold and complete in different cycles; i.e., they may be of different lengths. See Example A.10.

A.4.10   Sequence Length-Matching And (&&)

The *sequence length-matching and* operator specifies that two sequences must hold and complete in the same cycle; i.e., they must be of the same length. See Example A.11.

A.4.11   Sequence Or (|)

The *sequence or* operator specifies that one of two alternative sequences must hold (see Example A.12).

E X A M P L E   A.12

Sequence Or Operator

```
SERE ::= SERE | SERE
```

E X A M P L E   A.13

The until* Operators

```
FL_Property   ::=
              FL_Property until! FL_Property
            | FL_Property until FL_Property
            | FL_Property until!_ FL_Property
            | FL_Property until_ FL_Property
```

A.4.12   The until* Sequence Operators

The until* operators (until, until!, until!_, and until_) specify that a property holds until a second property holds. The until* operators provide another way to move forward, this time while putting a requirement on the cycles in which we are moving. See Example A.13.

The different flavors of this operator specify strong (until! and until!_) or weak (until and until_) operators. *Strong operators* require the terminating property to eventually occur, while *weak operators* do not. The inclusive operators (until_ and until!_) specify that the property must hold up to and including the cycle in which the terminating property holds, whereas the noninclusive operators (until and until!) require the property to hold up to, but not necessarily including, the cycle in which the terminating property holds.

A.4.13   The within Sequence Operator

The SERE within operator, shown in Example A.14, constructs a SERE in which the second SERE holds at the current cycle, and the

EXAMPLE A.14

The within Operator

```
SERE   ::=
   SERE within SERE
```

first SERE starts at or after the cycle in which the second starts and completes at or before the cycle in which the second completes.

A.4.14   The next* Operators

The next* operators allow us to be more specific about the timing; they take us forward 1 clock cycle. The next operator comes in both weak (next) and strong (next!) forms. If the number parameter is present, it indicates the cycle at which the property on the right-hand side holds. (See Example A.15.)

For further information on the remaining family of next* operators, please refer to the PSL LRM.

A.4.15   The eventually! Operator

While the next operator moves us forward exactly 1 cycle, the eventually! operator allows us to move forward without specifying exactly when to stop. This operator is a strong operator, which requires that the ending property or sequence actually occur. (See Example A.16.)

EXAMPLE A.15

The next* Operators

```
FL_Property   ::=
          next FL_Property
        | next! FL_Property
        | next [ Number ] ( FL_Property )
        | next! [ Number ] ( FL_Property )
```

EXAMPLE A.16

The eventually! Operator

```
FL_Property   ::=
          eventually! FL_Property
        | eventually! Sequence
```

EXAMPLE A.17

The before* Operators

```
FL_Property   ::=
          FL_Property before! FL_Property
        | FL_Property before  FL_Property
        | FL_Property before!_ FL_Property
        | FL_Property before_  FL_Property
```

A.4.16 The before* Operators

The before* operators provide an easy way to state that some signal must be asserted before some other signal. (See Example A.17.) The different flavors of this operator specify strong (before! and before!_) or weak (before and before_) operators. Strong operators require the ending condition to eventually occur, while weak operators do not. If the ending condition overlaps with the rightmost operand of the sequence, use the inclusive operators (before_ and before!_). Use the noninclusive operators (before and before!) to require the rightmost operand of the sequence to complete the cycle before the terminating condition.

A.4.17 The abort Operator

The abort operator provides a way to lift any future obligations of a property when some boolean condition is observed (see Example A.18).

The abort operator is reminiscent of the until operator, but there is an important difference. Both f abort b and f until b

EXAMPLE A.18

The abort Operator

```
FL_Property   ::=
        FL_Property abort Boolean
```

EXAMPLE A.19

The endpoint Declaration

```
Endpoint_Declaration   ::=
        endpoint Name [ ( Formal_Parameter_List ) ] DEF_SYM Sequence ;
```

specify that we should stop checking when b occurs. However, the abort operator removes future obligations of f, while the until operator does not.

A.4.18 The endpoint Declaration

An endpoint for a sequence is defined in PSL with a named endpoint declaration. The name of an endpoint cannot be the same name as other named PSL declarations (see Example A.19).

A.4.19 Suffix Implication Operators

A SERE is not a PSL property in and of itself. To use a SERE to build a PSL property, we link a SERE with another PSL property or with another SERE by using an implication operator. An implication operator can be read as whenever we have a sequenceA, we expect to see sequenceB (see Example A.20).

The strong implication operators specify that the rightmost sequence must complete, whereas the weak implication operators do not. The suffix implication specifies that the rightmost sequence begins on the cycle in which the leftmost sequence ends. The suffix next implication specifies that the rightmost sequence begins on the cycle after the leftmost sequence ends.

EXAMPLE A.20

Suffix Implication

```
FL_Property   ::=
        Sequence ( FL_Property )
      | Sequence |-> Sequence [ ! ]
      | Sequence |=> Sequence [ ! ]
```

EXAMPLE A.21

Logical Implication

```
FL_Property   ::=
        FL_Property -> FL_Property
```

A.4.20   Logical Implication Operator

The logical if implication operator specifies that if the leftmost property holds, then the rightmost property must hold (see Example A.21).

A.4.21   The always Temporal Operator

The always operator specifies one of the simplest temporal properties, which states that some boolean expression must hold at all times (see Example A.22).

A.4.22   The never Temporal Operator

The never operator allows us to specify an invariant property, which specifies conditions that must *never* hold. (See Example A.23.)

EXAMPLE A.22

The always Operator

```
FL_Property   ::=
        always FL_Property
```

EXAMPLE A.23

The never Operator

```
FL_Property   ::=
        never FL_Property
```

EXAMPLE A.24

Named Property

```
Property_Declaration   ::=
        property Name [ ( Formal_Parameter_List ) ] DEF_SYM Property ;
```

A.5 PSL PROPERTIES

A.5.1 Property Declaration

The building blocks of boolean expressions and sequences described in previous sections create PSL properties. Properties capture the temporal relationships between these building blocks. Properties are grouped using parentheses.

A.5.2 Named Properties

PSL allows us to name property definitions as shown in Example A.24. Note that DEF_SYM is = for Verilog and is for VHDL.

A.5.3 Property Clocking

PSL allows us to declare a default clock, explicitly declare a clock associated with a property, or declare that "clock cycle" and "next point in time" are equivalent. A clock expression is any boolean expression. A PSL property may refer to multiple clocks. (See Examples A.25 and A.26.)

A.5.4 The forall Property Replication

PSL provides an easy way to replicate properties that are the same except for specific parameters. Example A.27 shows the syntax for

EXAMPLE A.25

Default Clock Declaration

```
PSL_Declaration   ::=
      Clock_Declaration
Clock_Declaration   ::=
      default clock DEF_SYM Clock_Expression ;
```

EXAMPLE A.26

Clocked Property or SERE

```
SERE   ::=
         SERE @ Clock_Expression

FL_Property   ::=
         FL_Property @ Clock_Expression
```

EXAMPLE A.27

The forall Property Replication Syntax

```
Property   ::=
   Replicator Property
Replicator   ::=
   forall Name [ IndexRange ] in ValueSet :
IndexRange ::=
   LEFT_SYM finite_Range RIGHT_SYM
Flavor Macro LEFT_SYM =
   Verilog: [ / VHDL: ( / GDL: (
Flavor Macro RIGHT_SYM =
   Verilog: ] / VHDL: ) / GDL: )
ValueSet   ::=
   { ValueRange { , ValueRange } }
   | boolean
ValueRange ::=
   Value
 | finite_Range
Range ::=
  LowBound RANGE_SYM HighBound
```

the `forall` operator. Note the `RANGE_SYM` is : for Verilog and SystemVerilog and `to` for VHDL.

A.6 THE VERIFICATION LAYER

The *verification layer* tells the verification tools what to do with the properties described by the temporal layer. For example, the verification layer contains directives that tell a tool to assert a property (i.e., to verify that it holds) or to check that a specified sequence is covered by some test case.

A.6.1 The assert Directive

The `assert` directive verifies that a property holds. If the property does not hold, an error is raised (see Example A.28).

A.6.2 The assume Directive

The `assume` directive defines constraints to guide a verification tool (see Example A.29).

A.6.3 The cover Directive

The `cover` directive instructs the tool to indicate if a property has been exercised by the test suite or given constraints (see Example A.30).

EXAMPLE A.28

The assert Directive

```
Assert_Statement  : :=
        assert Property ;
```

EXAMPLE A.29

The assume Directive

```
Assume_Statement  : :=
        assume Property ;
```

E X A M P L E   A.30

The cover Directive

```
Cover_Statement   ::=
     cover Sequence ;
```

A.7 THE MODELING LAYER

The *modeling layer* models the behavior of design inputs (for tools such as formal verification tools that do not use test cases) and auxiliary hardware that is not part of the design but is needed for verification. This layer is, for the most part, outside the scope of this appendix. However, in this section we discuss a few useful functions that are defined by the modeling layer (see Example A.31).

A.7.1 prev()

The built-in function prev() takes an expression of any type as an argument and returns a previous value of that expression. With a single argument, the built-in function prev() gives the value of the expression in the previous cycle, with respect to the clock of its context. If a second argument is specified and has the value *i*, the

E X A M P L E   A.31

Built-in Functions

```
Built_In_Function_Call   ::=
    prev ( AnyType [ , Number ] )
  | next ( AnyType )
  | stable ( AnyType )
  | rose ( Bit )
  | fell ( Bit )
  | isunknown (BitVector)
  | countones (BitVector)
  | onehot (BitVector)
  | onehot0 (BitVector)
```

built-in function `prev()` gives the value of the expression in the
*i*th previous cycle, with respect to the clock of its context.

The clock context may be provided by the PSL property in which
the function call is nested, or by a relevant default clock declaration. If
the context does not specify a clock, the relevant clock is that corresponding to the granularity of time as seen by the verification tool.

A.7.2   next()

The built-in function `next()`   gives the value of a signal of any
type at the next cycle, with respect to the finest granularity of time
as seen by the verification tool. In contrast to the built-in functions
`prev()`, `stable()`, `rose()`, and `fell()`, the function `next()`
is not affected by the clock of its context.

A.7.3   stable()

The built-in function `stable()` takes an expression of any type as
an argument and produces a boolean result that is *true* if the argument's value is the same as it was at the previous cycle, with
respect to the clock of its context.

The clock context may be provided by the PSL property in
which the function call is nested or by a relevant default clock declaration. If the context does not specify a clock, the relevant clock is
that corresponding to the granularity of time as seen by the verification tool.

The function `stable()` can be expressed in terms of the
built-in function `prev()` as follows: `stable(e)` is equivalent to
the Verilog or SystemVerilog expression (`prev(e)` `==` `e`) and is
equivalent to the VHDL expression (`prev(e)` `=` `e`), where `e` is
any expression. The function `stable()` can be used anywhere a
boolean is required.

A.7.4   rose()

The built-in function `rose()` takes a bit expression as an argument and produces a boolean result that is *true* if the argument's

value is 1 at the current cycle and 0 at the previous cycle, with respect to the clock of its context; otherwise it is *false*.

The clock context may be provided by the PSL property in which the function call is nested or by a relevant default clock declaration. If the context does not specify a clock, the relevant clock is that corresponding to the granularity of time as seen by the verification tool.

The function `rose()` can be expressed in terms of the built-in function `prev()` as follows: `rose(b)` is equivalent to the Verilog or SystemVerilog expression `(prev(b)==1'b0 && b==1'b1)` and is equivalent to the VHDL expression `(prev(b)='0' and b='1')`, where b is a bit signal. The function `rose(b)` can be used anywhere a boolean is required.

A.7.5   fell()

The built-in function `fell()` takes a bit expression as an argument and produces a boolean result that is *true* if the argument's value is 0 at the current cycle and 1 at the previous cycle, with respect to the clock of its context; otherwise, it is *false*.

The clock context may be provided by the PSL property in which the function call is nested or by a relevant default clock declaration. If the context does not specify a clock, the relevant clock is that corresponding to the granularity of time as seen by the verification tool.

The function `fell()` can be expressed in terms of the built-in function `prev()` as follows: `fell(b)` is equivalent to the Verilog or SystemVerilog expression `(prev(b)==1'b1 && b==1'b0)` and is equivalent to the VHDL expression `(prev(b)='1' and b='0')`, where b is a bit signal. The function `fell(b)` can be used anywhere a boolean is required.

A.7.6   isunknown()

The built-in function `isunknown()` takes a bit vector as an argument. It returns *true* if the argument contains any bits that have "unknown" values (i.e., values other than 0 or 1); otherwise, it returns *false*.

Function isunknown() can be used anywhere a boolean is required.

A.7.7 countones()

The built-in function countones() takes a bit vector as an argument. It returns a count of the number of bits in the argument that have the value 1.

Bits that have unknown values are ignored.

A.7.8 onehot(), onehot0()

The built-in function onehot() takes a bit vector as an argument. It returns *true* if the argument contains exactly 1 bit with the value 1; otherwise, it returns *false*.

The built-in function onehot0() takes a bit vector as an argument. It returns *true* if the argument contains at most 1 bit with the value 1; otherwise, it returns *false*.

For either function, bits that have unknown values are ignored.

Functions onehot() and onehot0() can be used anywhere a boolean is required.

A.8 BNF

This section summarizes the syntax.

A.8.1 Metasyntax

The formal syntax described in this standard uses the following extended *Backus-Naur Form* (BNF).

1. The initial character of each word in a nonterminal is capitalized, e.g.,

   ```
   PSL_Statement
   ```

   A nonterminal can be either a single word or multiple words separated by underscores. When a multiple-word nonterminal containing underscores is referenced within the text (e.g., in a statement that describes the semantics of the corresponding syntax), the underscores are replaced with spaces.

2. Boldface words are used to denote reserved keywords, operators, and punctuation marks as a required part of the syntax, e.g.,

```
vunit  ( ;
```

3. The : : = operator separates the two parts of a BNF syntax definition. The syntax category appears to the left of this operator, and the syntax description appears to the right of the operator. For example, item 4 shows three options for a Vunit_Type.

4. A vertical bar separates alternative items (use one only) unless it appears in boldface, in which case it stands for itself, e.g.,

```
Vunit_Type   ::=   vunit | vprop |   vmode
```

5. Square brackets enclose optional items unless they appear in boldface, in which case they stand for themselves, e.g.,

```
Sequence_Declaration   ::=
    sequence Name [ ( Formal_Parameter_List ) ]
DEF_SYM Sequence ;
```

indicates Formal_Parameter_List is an optional syntax item for Sequence_Declaration, whereas

```
| Sequence [ * [ Range ] ]
```

indicates that the (outer) square brackets are part of the syntax, while Range is optional.

6. Braces enclose a repeated item unless it appears in bold-face, in which case it stands for itself. A repeated item may appear zero or more times; the repetitions occur from left to right as with an equivalent left-recursive rule. Thus, the following two rules are equivalent:

```
Formal_Parameter_List   ::=
Formal_Parameter { ; Formal_Parameter }
```

```
Formal_Parameter_List  ::=
Formal_Parameter | Formal_Parameter_List ;
Formal_Parameter
```

7. A comment in a production is preceded by a colon (:) unless it appears in boldface, in which case it stands for itself.

8. If the name of any category starts with an italicized part, it is equivalent to the category name without the italicized part. The italicized part is intended to convey some semantic information. For example, vunit_Name is equivalent to Name.

9. Flavor macros, containing embedded underscores, are shown in uppercase. These reflect the various HDLs that can be used within the PSL syntax and show the definition for each HDL. The general format is the term Flavor Macro, then the actual *macro name*, followed by the = operator, and, finally, the definition for each of the HDLs; e.g.,

```
Flavor Macro RANGE_SYM =
    SystemVerilog: : / Verilog: : / VHDL: to
/ GDL: / ..
```

shows the *range symbol* macro (RANGE_SYM). See Sec. 4.3.2 for further details about *flavor macros*.

The main text uses *italic* type when a term is being defined, and monospace font for examples and references to constants such as 0, 1, or x values.

A.8.2 Tokens

PSL syntax is defined in terms of primitive *tokens*, which are character sequences that act as distinct symbols in the language.

Each PSL keyword is a single token. Some keywords end in one or two nonalphabetic characters (! or _ or both). Those characters are part of the keyword, not separate tokens.

Each of the following character sequences is also a token:

[	]	(	)	{	}
,	;	:	..	=	:=
*	+	|->	|=>	<->	->
[*	[+]	[->	[=		
&&	&	||	|	!	
$	@	.	/		

Finally, for a given flavor, the tokens of the corresponding HDL are tokens of PSL.

A.8.3 HDL Dependencies

PSL depends upon the syntax and semantics of an underlying hardware description language. In particular, PSL syntax includes productions that refer to nonterminals in SystemVerilog, Verilog, VHDL, or GDL. PSL syntax also includes flavor macros that cause each flavor of PSL to match that of the underlying HDL for that flavor.

For SystemVerilog, the PSL syntax refers to the following nonterminals in the Accellera SystemVerilog version 3.1a syntax:

◆ module_or_generate_item_declaration
◆ module_or_generate_item
◆ list_of_variable_identifiers
◆ identifier
◆ expression
◆ constant_expression

For Verilog, the PSL syntax refers to the following nonterminals in the IEEE 1364-2001 Verilog syntax:

◆ module_or_generate_item_declaration
◆ module_or_generate_item
◆ list_of_variable_identifiers
◆ identifier
◆ expression
◆ constant_expression
◆ net_declaration
◆ reg_declaration
◆ integer_declaration

For VHDL, the PSL syntax refers to the following nonterminals in the IEEE 1076-1993 VHDL syntax:

+ block_declarative_item
+ concurrent_statement
+ design_unit
+ identifer
+ expression
+ entity_aspect

For GDL, the PSL syntax refers to the following nonterminals in the GDL syntax:

+ module_item_declaration
+ module_item
+ module_declaration
+ identifer
+ expression

Verilog Extensions

For the Verilog flavor, PSL extends the forms of declaration that can be used in the modeling layer by defining two additional forms of type declaration. PSL also adds another form of expression for both Verilog and VHDL flavors.

```
Extended_Verilog_Declaration  ::=
        Verilog_module_or_generate_item_declaration
      | Extended_Verilog_Type_Declaration

Extended_Verilog_Type_Declaration  ::=
        integer Integer_Range list_of_variable_identifiers
;
      | struct { Declaration_List }
list_of_variable_identifiers ;

Integer_Range  ::=
        ( constant_expression : constant_expression )

Declaration_List  ::=
        HDL_Variable_or_Net_Declaration {
        HDL_Variable_or_Net_Declaration }
```

```
HDL_Variable_or_Net_Declaration ::=
      net_declaration
    | reg_declaration
    | integer_declaration
```

Flavor Macros

```
Flavor Macro DEF_SYM  =
      SystemVerilog: = / Verilog: = / VHDL: is / GDL: :=

Flavor Macro RANGE_SYM  =
      SystemVerilog: : / Verilog: : / VHDL: to / GDL: ..

Flavor Macro AND_OP  =
      SystemVerilog: && / Verilog: && / VHDL: and / GDL: &

Flavor Macro OR_OP  =
      SystemVerilog: || / Verilog: || / VHDL: or / GDL: |

Flavor Macro NOT_OP  =
      SystemVerilog: ! / Verilog: ! / VHDL: not / GDL: !

Flavor Macro MIN_VAL  =
      SystemVerilog: 0 / Verilog: 0 / VHDL: 0 / GDL: null

Flavor Macro MAX_VAL  =
      SystemVerilog: $ / Verilog: inf / VHDL: inf / GDL: null

Flavor Macro HDL_EXPR =
      SystemVerilog: SystemVerilog_Expression
     / Verilog: Verilog_Expression
     / VHDL: Extended_VHDL_Expression
     / GDL: GDL_Expression

Flavor Macro HDL_CLK_EXPR =
      SystemVerilog: SystemVerilog_Event_Expression
     / Verilog: Verilog_Event_Expression
     / VHDL: VHDL_Expression
     / GDL: GDL_Expression

Flavor Macro HDL_UNIT =
      SystemVerilog: SystemVerilog_module_declaration
     / Verilog: Verilog_module_declaration
     / VHDL: VHDL_design_unit
     / GDL: GDL_module_declaration

Flavor Macro HDL_MOD_NAME =
```

```
        SystemVerilog: module_Name
      / Verilog: module_Name
      / VHDL: entity_aspect
      / GDL: module_Name

Flavor Macro HDL_DECL =
        SystemVerilog:
        SystemVerilog_module_or_generate_item_declaration
      / Verilog: Extended_Verilog_Declaration
      / VHDL: VHDL_block_declarative_item
      / GDL: GDL_module_item_declaration

Flavor Macro HDL_STMT =
        SystemVerilog: SystemVerilog_module_or_generate_item
      / Verilog: Verilog_module_or_generate_item
      / VHDL: VHDL_concurrent_statement
      / GDL: GDL_module_item

Flavor Macro HDL_RANGE =
        VHDL: range_attribute_name

Flavor Macro LEFT_SYM =
        SystemVerilog: [  / Verilog: [  / VHDL: (  / GDL: (

Flavor Macro RIGHT_SYM =
        SystemVerilog: ]  / Verilog: ]  / VHDL: )  / GDL: )
```

A.8.4 Syntax Productions

The rest of this section defines the PSL syntax.

Verification Units

```
PSL_Specification  ::=
      { Verification_Item }

Verification_Item  ::=
      HDL_UNIT | Verification_Unit

Verification_Unit  ::=
      Vunit_Type PSL_Identifier [ ( Hierarchical_HDL_Name ) ] {
         { Inherit_Spec }
         { VUnit_Item }
      }
```

```
Vunit_Type   ::=
        vunit  |  vprop  |  vmode


Name   ::=
          HDL_or_PSL_Identifier


Hierarchical_HDL_Name   ::=
        HDL_MOD_NAME { Path_Separator   instance_Name }


Path_Separator   ::=
          .  |  /


Inherit_Spec   ::=
          inherit vunit_Name { , vunit_Name } ;


VUnit_Item ::=
          HDL_DECL
        | HDL_STMT
        | PSL_Declaration
        | PSL_Directive
```

PSL Declarations

```
PSL_Declaration   ::=
        Property_Declaration
      | Sequence_Declaration
      | Endpoint_Declaration
      | Clock_Declaration


Property_Declaration   ::=
        property PSL_Identifier [ ( Formal_Parameter_List ) ]
        DEF_SYM Property ;


Formal_Parameter_List   ::=
        Formal_Parameter { ; Formal_Parameter }


Formal_Parameter   ::=
        Param_Type PSL_Identifier { , PSL_Identifier }


Param_Type   ::=
        const  |  boolean  |  property  |  sequence


Sequence_Declaration   ::=
        sequence PSL_Identifier [ ( Formal_Parameter_List ) ]
        DEF_SYM Sequence ;


Endpoint_Declaration   ::=
```

```
        endpoint PSL_Identifier [ ( Formal_Parameter_List ) ]
        DEF_SYM Sequence ;

Clock_Declaration   ::=
        default clock DEF_SYM Clock_Expression   ;

Clock_Expression ::=
        boolean_Name
      | boolean_Built_In_Function_Call
      | Endpoint_Instance
      | ( Boolean )
      | ( HDL_CLK_EXPR )

Actual_Parameter_List   ::=
        Actual_Parameter { , Actual_Parameter }

Actual_Parameter   ::=
        Number | Boolean | Property | Sequence
```

PSL Directives

```
PSL_Directive   ::=
        [ Label : ] Verification_Directive
Label   ::=
        PSL_Identifier

HDL_or_PSL_Identifier ::=
        SystemVerilog_Identifier
      | Verilog_Identifier
      | VHDL_Identifier
      | GDL_Identifier
      | PSL_Identifier

Verification_Directive   ::=
          Assert_Directive
        | Assume_Directive
        | Assume_Guarantee_Directive
        | Restrict_Directive
        | Restrict_Guarantee_Directive
        | Cover_Directive
        | Fairness_Statement

Assert_Directive   ::=
        assert Property [ report String ] ;
Assume_Directive   ::=
        assume Property ;
```

```
Assume_Guarantee_Directive  ::=
        assume_guarantee Property [ report String ] ;

Restrict_Directive ::=
        restrict Sequence ;

Restrict_Guarantee_Directive ::=
        restrict_guarantee Sequence [ report String ] ;

Cover_Directive  ::=
        cover Sequence  [ report String ] ;

Fairness_Statement  ::=
        fairness Boolean ;
      | strong fairness Boolean , Boolean ;
```

PSL Properties

```
Property  ::=
          Replicator Property
        | FL_Property
        | OBE_Property

Replicator  ::=
        forall PSL_Identifier [ Index_Range ]  in Value_Set :

Index_Range ::=
        LEFT_SYM finite_Range RIGHT_SYM
      | ( HDL_RANGE )

Value_Set  ::=
          { Value_Range { , Value_Range } }
        | boolean

Value_Range  ::=
        Value
      | FiniteRange

Value  ::=
          Boolean
         | Number
(see "Sequences" below)
FL_Property  ::=
        Boolean
      | ( FL_Property )
      | Sequence [ ! ]
      | property_Name [ ( Actual_Parameter_List ) ]
```

```
        | FL_Property @ Clock_Expression
        | FL_Property abort Boolean
   : Logical Operators :
        | NOT_OP FL_Property
        | FL_Property AND_OP FL_Property
        | FL_Property OR_OP FL_Property
        :
        | FL_Property -> FL_Property
        | FL_Property <-> FL_Property
   : Primitive LTL Operators :
        | X FL_Property
        | X! FL_Property
        | F FL_Property
        | G FL_Property
        | [ FL_Property U FL_Property ]
        | [ FL_Property W FL_Property ]
   : Simple Temporal Operators :
        | always FL_Property
        | never FL_Property
        | next FL_Property
        | next! FL_Property
        | eventually! FL_Property
        :
        | FL_Property until! FL_Property
        | FL_Property until FL_Property
        | FL_Property until!_ FL_Property
        | FL_Property until_ FL_Property
        :
        | FL_Property before! FL_Property
        | FL_Property before FL_Property
        | FL_Property before!_ FL_Property
        | FL_Property before_ FL_Property
   : Extended Next (Event) Operators :
        | X [ Number ] ( FL_Property )
        | X! [ Number ] ( FL_Property )
        | next [ Number ] ( FL_Property )
        | next! [ Number ] ( FL_Property )
        :(see "Sequences" below)
        | next_a [ finite_Range ] ( FL_Property )
        | next_a! [ finite_Range ] ( FL_Property )
        | next_e [ finite_Range ] ( FL_Property )
        | next_e! [ finite_Range ] ( FL_Property )
        :
        | next_event! ( Boolean ) ( FL_Property )
        | next_event ( Boolean ) ( FL_Property )
        | next_event! ( Boolean ) [ positive_Number ] (
   FL_Property )
```

```
        | next_event ( Boolean ) [ positive_Number ] (
FL_Property )
            :
        | next_event_a! ( Boolean ) [ finite_positive_Range ]
( FL_Property )
        | next_event_a ( Boolean ) [ finite_positive_Range ]
( FL_Property )
        | next_event_e! ( Boolean ) [ finite_positive_Range ]
( FL_Property )
        | next_event_e ( Boolean ) [ finite_positive_Range ]
( FL_Property )
: Operators on SEREs :
        | { Sequence } ( FL_Property )
        | Sequence |-> FL_Property
        | Sequence |=> FL_Property
```

Sequential Extended Regular Expressions (SEREs)

```
SERE ::=
        Boolean
      | Sequence
      | Sequence_Instance
      | SERE ; SERE
      | SERE : SERE
      | Compound_SERE


Compound_SERE ::=
        Repeated_SERE
      | Braced_SERE
      | Clocked_SERE
      | Compound_SERE | Compound_SERE
      | Compound_SERE & Compound_SERE
      | Compound_SERE && Compound_SERE
      | Compound_SERE within Compound_SERE
```

Sequences

```
Sequence ::=
        Sequence_Instance
      | Repeated_SERE
      | Braced_SERE
      | Clocked_SERE


Repeated_SERE ::=
        Boolean [* [ Count ] ]
      | Sequence [* [ Count ] ]
      | [* [ Count ] ]
```

```
            | Boolean [+]
            | Sequence [+]
            | [+]
            | Boolean [= Count ]
            | Boolean [-> [ positive_Count ] ]

Braced_SERE ::=
        { SERE }

Sequence_Instance ::=
        sequence_Name [ ( Actual_Parameter_List ) ]

Clocked_SERE ::=
        Braced_SERE @ Clock_Expression

Count ::=
          Number
        | Range

Range ::=
        Low_Bound RANGE_SYM High_Bound

Low_Bound ::=
          Number
        | MIN_VAL

High_Bound ::=
          Number
        | MAX_VAL
```

Forms of Expression

```
Any_Type  ::=
        HDL_or_PSL_Expression

Bit  ::=
        bit_HDL_or_PSL_Expression

Boolean  ::=
        boolean_HDL_or_PSL_Expression

BitVector  ::=
        bitvector_HDL_or_PSL_Expression

Number  ::=
        numeric_HDL_or_PSL_Expression
String  ::=
        string_HDL_or_PSL_Expression
```

```
HDL_or_PSL_Expression ::=
       HDL_Expression
     | PSL_Expression
     | Built_In_Function_Call
     | Union_Expression
     | Endpoint_Instance

HDL_Expression ::=
       HDL_EXPR

PSL_Expression ::=
       Boolean -> Boolean
     | Boolean <-> Boolean

Built_In_Function_Call ::=
       prev (Any_Type [ , Number ] )
     | next ( Any_Type )
     | stable ( Any_Type )
     | rose ( Bit )
     | fell ( Bit )
     | isunknown ( BitVector )
     | countones ( BitVector )
     | onehot ( BitVector )
     | onehot0 ( BitVector )

Union_Expression ::=
       Any_Type union Any_Type

Endpoint_Instance ::=
       endpoint_Name [ ( Actual_Parameter_List ) ]
```

Optional Branching Extension

```
OBE_Property ::=
       Boolean
     | ( OBE_Property )
     | property_Name [ ( Actual_Parameter_List ) ]

: Logical Operators :
     | NOT_OP OBE_Property
     | OBE_Property AND_OP OBE_Property
     | OBE_Property OR_OP OBE_Property
     | OBE_Property -> OBE_Property
     | OBE_Property <-> OBE_Property
: Universal Operators :
     | AX OBE_Property
```

```
                    |  AG OBE_Property
                    |  AF OBE_Property
                    |  A [ OBE_Property U OBE_Property ]
         : Existential Operators :
                    |  EX OBE_Property
                    |  EG OBE_Property
                    |  EF OBE_Property
                    |  E [ OBE_Property U OBE_Property ]
```

IEEE 1800 SystemVerilog Assertions

B.1 INTRODUCTION TO IEEE 1800 SYSTEMVERILOG

In this appendix, we discuss the proposed IEEE 1800 SystemVerilog Assertions (SVA) standard, which is a component of the SystemVerilog language donated to IEEE by Accellera. Note that the Accellera and IEEE standards for SystemVerilog are evolving. In this appendix, we describe the Accellera SystemVerilog 3.1a standard, published in June 2004. Hence, it is possible that subtle changes have occurred in the language standard from what we present in this chapter. However, this appendix provides an excellent source to understand many of the key concepts of the SystemVerilog assertion language, and it builds an excellent foundation of knowledge that would be useful for understanding future generations of the language. We encourage you to refer to the Accellera SystemVerilog LRM for the full standard of the language (as well as the new IEEE 1800 LRM when it is available).

The BNF described here follows a few conventions.

♦ Keywords are in bold.
♦ Required syntax characters are marked with single quotes.

Production names not found in this text are part of the remainder of SystemVerilog BNF.

B.2 SVA OPERATORS AND KEYWORDS

SVA introduces keywords that allow the user to specify sequences and properties. Tables B.1 to B.4 discuss each of the new SVA operators and keywords.

TABLE B.1

SVA Keywords

property	endproperty	sequence	endsequence
and	or	intersect	within
throughout	first_match	ended	matchedassert
assume	cover		

TABLE B.2

SVA Operators

Name	Operator	Description
Consecutive repetition	s1 [* N:M]	Repetition of s1 N, or between N and M, times.
Goto repetition	s1 [-> N:M]	Repetition of s1 N to M times in nonconsecutive cycles, ending on s1.
Nonconsecutive repetition	s1 [= N:M]	Repetition of s1 N to M times in nonconsecutive cycles, maybe ending on s1.
Temporal delay	## N ## [N:M]	Concatenation of two sequence elements.
not	**not** p1	Invert result of evaluation of the property.
and	s1 **and** s1 p1 **and** p2	Both sequence properties occur at some time.
Intersect	s1 **intersect** s2	Both sequences occur at the same time.

(Continued)

Name	Operator	Description
or	s1 **or** s2 p1 **or** p2	Either sequence property occurs.
Condition	**if** (expr) p1 **if** (expr) p1 **else** p2	Based on evaluation of expr, evaluate property p1 if *true*, or p2 if *false*.
Boolean until	b **throughout** s1	b must be true until sequence s1 completes (results in a match).
within	s1 **within** s2	s1 and s2 must occur. Lengths of s1 and s2 must follow s1 <= s2.
ended	s1.**ended**	Sequence s1 matched (ended) at this time.
matched (from different clock domain)	s1.**matched**	Sequence s1 (on another clock) ended at this time.
first match	**first_match**(s1)	First occurrence of sequence, rest ignored.
Overlapping implication	s1 |-> p1	If s1 occurs, p1 (starting this cycle) must occur; else *true*.
Nonoverlapping implication	s1 |=> p1	If s1 occurs, p1 (starting next cycle) must occur; else *true*.

TABLE B.3

SVA System Functions

$rose	$fell	$stable
$countones	$onehot	$onehot0
$isunknown	$past	$sampled

B.3 SEQUENCE AND PROPERTY

The sequence operators defined for SystemVerilog allow us to compose expressions into temporal sequences. These sequences are the building blocks of properties and concurrent assertions. The first four allow us to construct sequences, while the remaining operators allow us to specify operations on a sequence as well (such as compose a complex sequence from simpler sequences).

TABLE B.4

SVA Operator Precedence

Operator (Sequence or Property)	Associativity
Repetition (consecutive, nonconsecutive, goto)	
Delay (##)	Left
Boolean until (**throughout**)	Right
within	Left
Intersection	Left
not (property)	
and (sequence and property)	Left
or (sequence and property)	Left
if..else	
Implication (overlapping, nonoverlapping)	Right

EXAMPLE B.1

Construction of Sequences with Temporal Delay

```
sequence_expr ::= [ cycle_delay_range ]
      sequence_expr { sequence_expr cycle_delay_range }

cycle_delay_range ::=
   ##    constant_expression
   ## '[' constant_expression ':' constant_expression ']'
   ## '[' constant_expression ':' '$'  ']' — Infinite range.
```

B.3.1 Temporal Delay

The temporal delay operator ## constructs larger sequences by combining smaller sequences and expressions (see Example B.1).

A temporal delay may begin a sequence. The range may be a single constant amount or a range of time. All times may be used to match the following sequence. The range is interpreted as follows:

```
##0 a        same as    (a)
##1 a        same as    (1'b1 ##1 a)
##[0:1] a    same as    a or (1'b1 ##1 a)
```

When the symbol $ is used, the range is infinite, e.g.,

```
## [0:$]  a
```

The right-side sequence of a concatenation is expected on the following cycle unless the delay has the value 0. Each subsequent element of a concatenation further advances the time 1 cycle. Examples include these:

`1'b1 ##1 a`	means a must be *true* in cycle 1
`1'b1 ##1 a ##1 b`	means a must be *true* in cycle 1, b must be true in cycle 2
`a ##[0:2] b`	same as `a & b` or `(a ##1 b)` or `(a ##1 1 ##1 b)`
`a ##[2:3] b`	same as `(a ##1 1 ##1 b)` or `(a ##1 1 ##1 1 ##1 b)`

Note: When a range is used, it is possible for the sequence to match with each value within the range. Thus, we must take into account that there may be multiple possible matches when we write sequences with a range. See Sec. B.3.12, "The `first_match`," for more information.

B.3.2 Consecutive Repetition

Consecutive repetition of a sequence applies to a single element or a sequence expression (see Example B.2).

When you are using a repetition, the element or sequence must occur starting in the next cycle for each repetition expected. Examples are

`a [* 0] ##1 b`	same as	`(b)`
`a [* 2] ##1 b`	same as	`(a ##1 a ##1 b)`
`a [* 1:2] ##1 b`	same as	`(a ##1 b)` or `(a ##1 a ##1 b)`
`(a ##1 b) [* 2]`	same as	`(a ##1 b ##1 a ##1 b)`

Note: When a range is used, it is possible for the sequence to match with each value within the range. Thus, we must take into account that there may be multiple possible matches when we write sequences with a range. See Sec. B.3.12, "The `first_match`," for more information.

EXAMPLE B.2

Consecutive Repetition of a Sequence

```
sequence_expr ::=
      sequence_expr          '[' '*' const_range_expression ']'
    | expression_or_dist   '[' '*' const_range_expression ']'
const_range_expression ::=
      constant_range_expression
    | constant_range_expression ':' constant_range_expression
    | constant_range_expression ':' '$'
```

EXAMPLE B.3

Goto Repetition of a Sequence

```
sequence_expr ::=
      expression_or_dist '[' '->' const_range_expression ']'

const_range_expression ::=   constant_expression
    | constant_expression ':' constant_expression
    | constant_expression ':' '$'
```

B.3.3 Goto Repetition

The goto repetition operator specifies finitely many iterative matches of the boolean expression operand, with a delay of one or more clock ticks from one match of the operand to the next successive match and no match of the operand strictly in between. The overall repetition sequence matches at the last iterative match of the operand. See Example B.3.

Goto repetition is defined in terms of the other operators as

```
s [-> min:max] ::= (!s[*0:$] ##1 s) [* min:max]
```

Examples are

```
a[->0] ##1 b      same as   (b)
a[->1] ##1 b      same as   (!a [*0:$] ##1 a ##1 b)
a[->2] ##1 b      same as   (!a [*0:$] ##1 a ##1 !a
                            [*0:$] ##1 a ##1 b)
```

EXAMPLE B.4

Nonconsecutive Repetition of a Sequence

```
sequence_expr ::=
    expression_or_dist '[' '=' const_range_expression ']'

const_range_expression ::=  constant_range_expression
  | constant_range_expression ':' constant_range_expression
  | constant_range_expression ':' '$'
```

B.3.4   Nonconsecutive Repetition

Nonconsecutive repetition, like *goto repetition,* allows for space between the repetition of the expression. At the end of the repetition, however, there can be additional clock ticks after the repeated expression (where the boolean expression does not match). See Example B.4.

Nonconsecutive repetition is defined in terms of the other operators as

```
s [= min:max] ::= ((!s[*0:$] ##1 s ) [* min:max ])
                   ##1 !s[*0:$]
```

Examples are

```
a [=0] ##1 b     same as   (b)
a [=1] ##1 b     same as   (!a [*0:$] ##1 a ##1 !a
                            [*0:$] ##1 b)
a [=2] ##1 b     same as   (!a [*0:$] ##1 a ##1 !a
                            [*0:$] ##1 a ##1 !a
                            [*0:$] ##1 b)
```

B.3.5   Sequence and Property and

Both properties and sequences can be operands of the and operator. The operation on two sequences produces a match once both sequences produce a match (the endpoint may not match). A match occurs until the endpoint of the longer sequence (provided the shorter sequence produces one match). See Example B.5.

EXAMPLE B.5

Sequence and Property and (Non-Matching-Length)

```
sequence_expr ::= sequence_expr and sequence_expr
```

This operation, for sequences, is defined in terms of the next operator (intersect) as

```
s and t ::= ( (s ##1 1 [*0:$]) intersect t) or
            (s intersect (t ##1 1 [*0:$])))
```

Examples of sequence and'ing are [() means no match]

```
(a ##1 b) and ()        same as      ()
(a ##1 b) and           same as      (a & c ##1 b & d)
(c ##1 d)
(a ##[1:2] b)           same as      (a & c ##1 b ##1 1
and (c ##3 d)                        ##1 d) or (a & c ##1
                                     1 ##1 b ##1 d)
```

Note: It is possible for this sequence operation to match with each value of a range or repeat in the operands. Thus, we must take into account that there may be multiple possible matches when we use this operator. See Sec. B.3.12, "The first_match," for more information.

Properties can be combined using this operator to produce a match if both properties evaluate to *true*. Only one match will be produced when the operands are properties. A sequence is implicitly transformed into a property through the application of first_match to the sequence. See Sec. B.3.13, "Property Implication," for additional information.

B.3.6 Sequence Intersection

An *intersection* of two sequences is like an and of two sequences (both sequences produce a match). This operator also requires the lengths of the sequences to match. That is, the match point of both sequences must be the same time. With multiple matches of each sequence, a match occurs each time both sequences produce a match. See Example B.6.

EXAMPLE B.6

Sequence and (Matching-Length)

```
sequence_expr ::= sequence_expr intersect sequence_expr
```

Examples of a sequence `intersect` are [() means no match]

(1) **intersect** ()	same as	()
##1 a **intersect** ##2 b	same as	()
##2 a **intersect** ##2 b	match if	(##2 (a & b))
## [1:2] a **intersect** ## [2:3] b	match if	(1 ##1 a&b) or (1 ##1 a&b ##1 b)
## [1:3] a **intersect** ## [1:3] b	match if	(##1 a&b) or (##2 a&b) or (##3 a&b)

Note: It is possible for this operation to match with each value of a range or repeat in the operands. Thus, we must take into account that there may be multiple possible matches when we use this operator. See Sec. B.3.12, "The `first_match`," for more information.

B.3.7   Sequence and Property or

For sequences, a match on either sequence results in a match for the operation (see Example B.7). Examples of sequence `or` are

() **or** ()	same as	()
(## 2 a **or** ## [1:2] b)	match if	(b) or (##1 b) or (## 2 a) or (##2 b)

Note: It is possible for this operation to match with each value of a range or repeat in the operands. Thus, we must take into account that there may be multiple possible matches when we use this operator. See Sec. B.3.12, "The `first_match`," for more information.

Properties can be combined using this operator to produce a match if one property evaluates to *true*. Only one match will be produced when the operands are properties. A sequence is implicitly

EXAMPLE B.7

Sequence or

```
sequence_expr ::= sequence_expr or sequence_expr
```

EXAMPLE B.8

The Boolean throughout Sequence

```
sequence_expr ::= expression_or_dist throughout sequence_expr
```

transformed into a property through the application of
`first_match` to the sequence. See Sec. B.3.13, "Property
Implications," for additional information.

B.3.8 Boolean until (throughout)

This operator matches a boolean value throughout a sequence. The
operator produces a match if the sequence matches and the boolean
expression is *true* until the end of the sequence. See Example B.8.

The `throughout` operator is defined in terms of the
`intersect` sequence operators. Its definition is

```
(b throughout s) ::= ( b [*0:$] intersect s)
```

Examples include

```
0 throughout (1)       same as   ()
1 throughout ##1 a     same as   ##1 a
a throughout ##2 b     same as   (a ##1 a & b)
a throughout           same as   (a&b ##1 a ##1 a &c) or
(b ##[1:3] c)                    (a&b ##1 a ##1 a ##1 a
                                 &c) or (a&b ##1 a ##1 a
                                 ##1 a ##1 a &c)
```

Note: It is possible for this operation to match with each value of
a range or repeat in the operands. Thus, we must take into account
that there may be multiple possible matches when we use this operator. See Sec. B.3.12, "The `first_match`," for more information.

EXAMPLE B.9

The within Sequence

```
sequence_expr ::= sequence_expr within sequence_expr
```

B.3.9 The within Sequence

The within operator determines if one sequence matches within the length of another sequence (see Example B.9). The within operator is defined in terms of the intersect sequence operators. Its definition is

```
s1 within s2 ::= ((1 [*0:$] ##1 s1 ##1 1 [*0:$])
                    intersect s2)
```

This means that s1 may start the same time as s2 or later and may end earlier or at the same time as s2. Examples are [() means no match]

() **within** (1)	same as	()	
(1) **within** ()	same as	()	
(a) **within** ## [1:2] b	same as	(a&b) or (b ##1 a)	
		or (a ##1 b)	
		or (##1 a&b)	

Note: It is possible for this operation to match with each value of a range or repeat in the operands. Thus, we must take into account that there may be multiple possible matches when we use this operator. See Sec. B.3.12, "The first_match," for more information.

B.3.10 The Method ended

The method ended returns true at the end of the sequence. This is in contrast to matching a sequence from the beginning time point, which is obtained when we use only the sequence name (see Example B.10). Examples include

```
sequence a1;
   @(posedge clk) (c ##1 b ##1 d);
endsequence
```

E X A M P L E   B.10

The ended Sequence

```
expression ::= sequence_identifier ended

sequence_identifier ::= (identifier of type sequence)
```

```
(a ##1   a1.ended)    same as    (c ##1 b & a ##1 d)
(a ##2   a1.ended)    same as    (c &a ## b ##1 d)
```

Note the position of 'a' relative to the end of sequence 'a1' (term 'd'). Compare this with the following sequence where 'a' occurs before 'a1'.

```
(a ##1   a1)      same as    (a ##1 c ##1 b ##1 d)
```

B.3.11   The Method matched

The method matched operates similarly to the ended method; however, this method is used when the sequence of the method call uses a different clock from the sequence being called (see Example B.11).

B.3.12   The first_match

The first_match operator returns only the first match from a sequence. The remaining are suppressed (see Example B.12).
Examples of first_match are

```
first_match (1 [*1:5])    same as    (1)
first_match (##[0:4] 1)    same as    (1)
```

E X A M P L E   B.11

The matched Sequence

```
expression ::= sequence_identifier '.' matched

sequence_identifier ::= (identifier of type sequence)
```

EXAMPLE B.12

The first_match

```
sequence_expr ::= first_match sequence_expr
```

first_match (##[0:1] a)	same as	(a) or (!a ##1 a)
first_match	same as	(s1 **intersect**
(s1 **intersect** s2)		**first_match**(s2)
first_match	same as	(b **throughout**
(b **throughout** s1)		**first_match**(s1))
first_match	same as	(s1 **within**
(s1 **within** s2)		**first_match** (s2))
first_match(expression)	same as	(expression [->1])

Note: Use of a range on the delay operators or a range on the repetition operators can cause multiple matches. Use of first_match can be helpful to suppress the subsequent matches. These additional matches may cause a false firing that is solved with this operator.

B.3.13 Property Implication

As a convenience, there are two forms of implication: *overlapping* and *nonoverlapping*. The overlap occurs between the final cycle of the left-hand side (the antecedent) and the starting cycle of the right-hand side (the consequent) operands. For the overlapping form, the consequent starts on the current cycle (that the antecedent matched). The nonoverlapping form has the consequent start the subsequent cycle. Implication is similar in concept to an if() statement.

Implication uses the antecedent as a test condition to determine whether the consequent should be evaluated or the operation should (vacuously) return a *true* result. The consequent is a property, allowing for negations, implication, and other property operations. See Examples B.13 and B.14.

E X A M P L E   B.13

Overlapping Implication

```
property_expr ::= sequence_expr '|->' property_expr
```

E X A M P L E   B.14

Nonoverlapping (Suffix) Implication

```
property_expr ::= sequence_expr '|=>' property_expr
```

Examples include

a |-> b	same as	a ? b : 1'b1	
(a ##1 b) |-> (c)	same as	(a ##1 b) ? c : 1'b1	
(a ##1 b) |=>	same as	(a ##1 b) |->	
(c ##1 d)		##1 c ##1 d)	

When the consequent property[†] is simply a sequence, the first match of the sequence is sufficient for the property to return *true*. This is equivalent to writing

```
antecedent |-> first_match(property_is_a_sequence)
```

Thus, a property will produce a single match or no match, regardless of the operators used.

B.3.14 Conditional Property Selection

Properties can be conditionally selected using the `if...else` operator (see Example B.15). The *conditional property selection* operator operates similarly to the procedural `if()` statement. The property evaluates to *true* if the (boolean) expression is *true* and `property_expr1` evaluates to *true*, or the expression is *false* and `property_expr2` evaluates to *true*.

[†]Actually any property that is only a sequence (does not contain the property operations) is satisfied on the first match, equivalent to implicitly using the `first_match()` operator.

EXAMPLE B.15

Conditional Selection Using if...else

```
property_expr ::= if (expression) property_expr1
    |    if (expression) property_expr1 else property_expr2
```

Examples include

```
A |=> first_match( ##[1:10] B | C )
  |-> if (B)   D ##[1:10] F
      else     G [*2]
```

The first line of the property identifies that sequence B after A or C after A.[†] The second line uses the conditional selection operator to choose the next property based on which sequence was detected. If B was found the property D ## [1:10] F is expected to match. If B was not found (implying C was found), the property G [*2] is expected to match.

B.4 PROPERTY DECLARATIONS

SystemVerilog allows for declarations of both sequences and properties. A property differs from a sequence in that it contains a clock specification for the entire property, an asynchronous term to disable property evaluations, and additional operators that can be used. Properties allow the operators negation, if..else, and implication operations. (See Example B.16.)

Properties can be complete definitions useful with other properties, or they can be used for verification as an assertion, assumption, or a coverage point. Properties can also contain parameters to be specified when they are used in these other contexts. Examples include

```
property req_t1_start;
    @(posedge clk)   req && req_tag == t1;
endproperty
```

[†]The first_match operator is used to prevent a subsequent shorter match from incorrectly completing the property. We could have used the sequence (b | c) [->1] if a time limit were not required.

EXAMPLE B.16

Property Declaration

```
property_declaration ::=
        property property_identifier [ property_formal_list ] ';'
                { property_decl_item }
                property_spec ';'
        endproperty [ ':' property_identifier ]

property_formal_list ::= '(' formal_list_item { ','
formal_list_item } ')'
formal_list_item       ::=  formal_identifier [ '=' actual_arg_expr ]
actual_arg_expr   ::=  expression | identifier | event_control | '$'

property_decl_item ::=  sequence_declaration
        | list_of_variable_identifiers_or_assignments

property_spec ::=   [ event_control ]
   [ disable iff '(' expression ')' ]   property_expr

property_expr ::=
     sequence_expr
   | event_control   property_expr
   | '('   property_expr ')'
   | not   property_expr
   | property_expr  or    property_expr
   | property_expr  and   property_expr
   | sequence_expr  |-> property_expr
   | sequence_expr  |=> property_expr
   | if '(' expression ')' property_expr
        [ else property_expr ]
   | property_instance

event_control    ::= '@' event_identifier
   | '@' '(' event_expression ')'
```

```
        property illegal_op;
            @(posedge clk)  not req && cmd == 4;
        not @(posedge clk)   req && cmd == 4;
        endproperty
        property starts_at(start, grant, reset_n);
```

EXAMPLE B.17

Sequence Declaration

```
sequence_declaration ::=
  sequence sequence_identifier [sequence_formal_list ] ';'
    { assertion_variable_declaration }
    sequence_expr ';'
  endsequence [ ':' sequence_identifier ]

sequence_formal_list ::=
  '(' formal_list_item { ',' formal_list_item } ')'

assertion_variable_declaration ::=
  data_type  list_of_variable_declarations ';'
```

```
        disable iff  (~reset_n)     // asynch reset of property.
        @(posedge clk) (grant[->1] ##1 grant & start);
    endproperty
```

Properties may reference other properties in their definition. They may even reference themselves recursively. Properties may also be written utilizing multiple clocks to define a sequence that transitions from one clock to another as it matches the elements of the sequence. See the SystemVerilog LRM for additional details.

Sequences are also declared. They use syntax similar to properties. See Example B.17.

They can be defined within properties and as separate elements. They also can be declared with parameters that are specified when used in other contexts. These elements, coupled with the following directives, allow one to define and utilize properties to follow an assertion-based design methodology. Examples include

```
sequence op_retry;
    (req ##1 retry);
endsequence
sequence cache_fill(req, done, fill);
    (req ##1 done [-> 1] ##1 fill [-> 1]);
endsequence
```

B.4.1   Sequence Composition

Sequence creation can use all the operators defined above, except the property operators implication, conditional selection (`if...else`), and negation. Example B.18 shows the BNF.

B.5   THE ASSERT, ASSUME, AND COVER STATEMENTS

Property directives define how to use properties (and sequences) for specific works. These statements utilize all the elements above to define how they are to be used for a given design (see Example B.19).

As an *assertion*, properties are evaluated, and when they do not match the desirable sequence, they fail and produce an error message by default (or they execute the **else** statement set). When they match a sequence, the first (pass) statement set is executed (like an **if()** statement).

As an *assumption*, the properties are expected to hold true during simulation. For formal property tools, the assumptions are also expected to hold true during a proof. The formal tool is not required to verify all assumptions are true, though, allowing one to specify assumptions restricting the logic to a subset of all legal input values. This restriction may be to simplify proofs during development.

As a *cover* directive, properties are evaluated, and when they succeed, they execute the first (pass) statement set. If they fail to match a sequence, they execute the **else** statement set, if any.

B.6   SYSTEM FUNCTIONS

Assertions are commonly used to evaluate certain specific characteristics of a design implementation, such as whether a particular signal is one-hot. The following system functions are included to facilitate such common assertion functionality.

- ♦ `$onehot (<expression>)` returns *true* if one and only 1 bit of expression is high.
- ♦ `$onehot0 (<expression>)` returns *true* if at most 1 bit of expression is high.

E X A M P L E   B.18

Sequence Specification

```
cycle_delay_range ::=
    '##' constant_expression
  | '##' '[' const_range_expression ']'

const_range_expression ::=
    constant_expression : constant_expression
  | constant_expression : '$'

sequence_expr ::= [ cycle_delay_range ]
        sequence_expr { cycle_delay_range sequence_expr }
  | expression_or_dist [boolean_repeat]
  | expression_or_dist { ',' sequence_match_item }
[boolean_repeat]
  | sequence_instance [ consecutive_repeat ]
  | sequence_expr { ',' sequence_match_item }
[consecutive_repeat]
  | event_control sequence_expr
  | '(' sequence_expr ')'
  | sequence_expr and        sequence_expr
  | sequence_expr or         sequence_expr
  | sequence_expr intersect  sequence_expr
  | sequence_expr within     sequence_expr
  | expression     throughout sequence_expr

sequence_match_item ::= variable_assignment | subroutine_call

expression_or_dist ::= expression
  | expression dist '{' dist_list '}'

boolean_repeat ::= consecutive_repeat
  | nonconsecutive_repeat
  | goto_repeat

consecutive_repeat    ::=
            '[' '*' const_range_expression ']'
nonconsecutive_repeat ::=
            '[' '=' const_range_expression ']'
goto_repeat           ::=
            '[' '->' const_range_expression ']'
```

EXAMPLE B.19

Property Directives

```
immediate_assert_statement ::=
    assert ( expression ) action_block


concurrent_assert_statement ::=
 assert property '(' property_spec ')' action_block


concurrent_assume_statement ::=
 assume property '(' property_spec ')' ';'


concurrent_cover_statement ::=
 cover property '(' property_spec ')' statement_or_null


action_block ::=
    statement    [ else statement_or_null ]
  | [statement_or_null] else statement_or_null


statement_or_null ::=
    statement    |    ';'
```

- ◆ $isunknown (<expression>) returns *true* if any bit of the expression is 'X' or 'Z'. This is equivalent to

  ^ <expression> === 1'bx

- ◆ $stable (<expression>, <clocking event>) returns *true* if the previous value of the expression is the same as the current value of the expression.

- ◆ $rose (<expression>, <clocking event>) returns *true* if the expression was previously zero and the current value is nonzero. If the expression length is more than 1 bit, then only bit 0 is used to determine a positive edge.

- ◆ $fell (<expression>, <clocking event>) returns *true* if the expression was previously nonzero and the current value is zero. If the expression length is more than 1 bit, then only bit 0 is used to determine a negative edge.

EXAMPLE B.20

The $past Equivalent Code

```
assign old_data =
        $past(data, 2, ld_done, posedge clk);

always @(posedge clk)
  if (ld_done)
    {old_data, prev_data} <= { prev_data, data};
```

All the above system functions have a return type of bit. A return value of 1'b1 indicates *true*, and a return value of 1'b0 indicates *false*. These following system functions return a value equivalent to the length of the first (or only) expression.

+ $past (<expression>, <num cycles>, <clock gate>, <clocking event>) returns the value of the expression from *num cycles* ago. If the value did not exist, `bx is returned. The clock gate expression, if specified, causes the evaluation of the clock to be dependent on its asserted value. The clocking event, if specified, explicitly defines the sampling clock. For example the $past call is equivalent to the code shown in Example B.20.

+ $sampled (<expression>, <clocking event>) returns the sampled value of the expression. If the value did not exist, `bx is returned. The clocking event is implicitly inferred unless explicitly specified as the second expression.

+ $countones (<expression>) returns the number of bits active high (for example, 1'b1).

B.7 SYSTEM TASKS

SystemVerilog has defined several severity system tasks for use in reporting messages with a common message. These system tasks are defined as follows:

```
$fatal(finish_num [, message
        {, message_argument } ] );
```

This system task reports the error message provided and terminates the simulation with the `finish_num` error code. This system task is best used to report fatal errors related to testbench/OS system failures (e.g., cannot open, read, or write to files) The message and argument format is the same as the `$display()` system task.

> `$error(` message `{,` message_argument `}` `)`;

This system task reports the error message as a runtime error in a tool-specific manner. However, it provides the following information:

* Severity of the message
* File and line number of the system task call
* Hierarchical name of the scope or assertion or property
* Simulation time of the failure

> `$warning(`message `{,` message_argument `}` `)`;

This system task reports the warning message as a run-time warning in format similar to `$error` and with similar information.

> `$info(`message `{,` message_argument `}` `)`;

This system task reports the informational message in a format similar to `$error` and with similar information.

> `$asserton(`levels, `[`
> list_of_modules_or_assertions])

This system task will reenable assertion and coverage statements. This allows sequences and properties to match elements. If a `level` of 0 is given, all statements in the design are affected. If a list of *modules* is given, then that module and modules instantiated to a depth of the `level` parameter are affected. If specifically named assertion statements are listed, then they are affected.

> `$assertkill(`levels, `[`
> list_of_modules_or_assertions])

This system task stops the execution of all assertion and cover statements. These statements will not begin matching until reenabled with **$asserton()**. Use the arguments in the same way as `$asserton` uses them.

> `$assertoff(`levels, `[`
> list_of_modules_or_assertions])

This system task prevents matching of assertion and cover statements. Sequences and properties in the progress of matching sequences will continue. Assertion and cover statements will not begin matching again until reenabled with **$asserton()**. Use the arguments in the same way as $asserton uses them.

B.8 BNF

The SVA BNF represented here is the property specification subset. This subset resides within a module context. It also utilizes the expression BNF subset as part of its expressions. For the complete SVA BNF, refer to the Accellera (or IEEE) LRM.

B.8.1 Use of Assertions BNF

```
concurrent_assertion_item::=
            | concurrent_assert_statement
            | concurrent_assume_statement
            | concurrent_cover_statement
procedural_assertion_item::=
            immediate_assert_statement
            | concurrent_assert_statement
            | concurrent_cover_statement
```

B.8.2 Assertion Statements

```
immediate_assert_statement ::=
    assert ( expression ) action_block
concurrent_assert_statement ::=
    assert property '(' property_spec ')' action_block
concurrent_assume_statement ::=
        assume property '(' property_spec ')'
concurrent_cover_statement ::=
        cover property '(' property_spec ')'
statement_or_null
action_block        ::=
        statement                [ else statement_or_null ]
        | [statement_or_null] else statement_or_null
```

```
statement_or_null ::=
        statement    |   ';'
```

B.8.3   Property and Sequence Declarations

```
property_declaration ::=
  property property_identifier [ property_formal_list ] ';'
     { assertion_variable_declaration }
                property_spec ';'
   endproperty [ ':' property_identifier ]


property_formal_list ::=
         '(' formal_list_item { ',' formal_list_item } ')'
formal_list_item ::= formal_identifier [ '=' actual_arg_expr ]
actual_arg_expr ::=  expression | identifier |  event_control


property_decl_item ::=  sequence_declaration
       | list_of_variable_identifiers_or_assignments
sequence_declaration ::=
  sequence sequence_identifier [sequence_formal_list ] ';'
     { assertion_variable_declaration }
       sequence_expr ';'
   endsequence [ ':' sequence_identifier ]
sequence_formal_list ::=
       '(' formal_list_item { ',' formal_list_item } ')'
assertion_variable_declaration ::=
          data_type   list_of variable_identifiers
```

B.8.4   Property Construction

```
property_spec ::=  [ event_control ]
  [ disable iff '(' expression ')' ] multi_clock_property_expr


property_expr ::=
    sequence_expr
  | event_control  property_expr
  | '(' property_expr ')'
  | not  property_expr
  | property_expr  or   property_expr
  | property_expr  and  property_expr
  | sequence_expr |-> property_expr
  | sequence_expr |=> property_expr
  | if '(' expression ')' property_expr
```

```
    [ else property_expr ]
  | property_instance


event_control   ::= '@' event_identifier
      | '@' '(' event_expression ')'
```

B.8.5 Sequence Construction

```
cycle_delay_range ::=
    '##' constant_expression
  | '##' '[' const_range_expression ']'


const_range_expression ::= constant_expression : '$'
  |   constant_expression : constant_expression


sequence_expr ::= [ cycle_delay_range ]
        sequence_expr { cycle_delay_range sequence_expr }
  | expression_or_dist [boolean_repeat]
  | expression_or_dist { ',' sequence_match_item } [boolean_repeat]
  | sequence_instance [ consecutive_repeat ]
  | sequence_expr { ',' sequence_match_item }
[consecutive_repeat]
  | event_control  sequence_expr
  | '(' sequence_expr ')'
  | sequence_expr and         sequence_expr
  | sequence_expr or           sequence_expr
  | sequence_expr intersect  sequence_expr
  | sequence_expr within       sequence_expr
  | expression_or_dist    throughout  sequence_expr


sequence_match_item ::= variable_assignment | subroutine_call
expression_or_dist ::= expression
      | expression dist '{' dist_list '}'


boolean_repeat ::= consecutive_repeat
  |  nonconsecutive_repeat
  |  goto_repeat'


consecutive_repeat   ::=
  '[' '*'   const_range_expression ']'
nonconsecutive_repeat ::=
  '['   '=' const_range_expression ']'
goto_repeat           ::=
  '[' '->' const_range_expression ']'
```

BIBLIOGRAPHY

[Accellera OVL 2003] Accellera proposed standard *Open Verification Library Users Reference Manual*, 2003.

[Accellera FVTC 2004] Accellera Formal Verification Technical Committee, www.eda.org/vfv.

[Accellera PSL-1.1 2004] Accellera proposed standard Property Specification Language (PSL) 1.1, March 2004.

[Accellera SystemVerilog-3.1a 2004] Accellera proposed standard SystemVerilog 3.1a, March 2004.

[AMBA-2.0 1999] AMBA™ Specification, Revision 2.0, ARM Limited, 1999.

[Bening and Foster 2001] L. Bening and H. Foster, *Principles of Verifiable RTL Design*, Kluwer Academic Publishers, 2001.

[Bergeron 2003] J. Bergeron, *Writing Testbenches: Functional Verification of HDL Models*, Kluwer Academic Publishers, 2003.

[Berman and Trevillyan 1989] C. L. Berman and L. H. Trevillyan, "Functional Comparison of Logic Designs for VLSI Circuits," *Proc. Int'l Conf. Computer-Aided Design* (ICCAD 89), IEEE CS Press, Los Alamitos, Calif., 1989, pp. 456–459.

[Brand 1993] D. Brand, "Verification of Large Synthesis Designs," *Proc. Int'l Conf. Computer-Aided Design* (ICCAD 93), IEEE CS Press, Los Alamitos, Calif., 1993, pp. 534–537.

[Bryant 1986] R. E. Bryant, "Graph-Based Algorithms for Boolean Function Manipulation," *IEEE Transactions on Computers* C-35(8):677–691, August 1986.

[Chockler et al. 2001] H. Chockler, O. Kupferman, R. P. Kurshan, and M. Y. Vardi, "A Practical Approach to Coverage in Model Checking," in *Proc. 13th CAV*, LNCS 2102, pp. 66–78, 2001.

[Clarke and Emerson 1981] E. Clarke and E. A. Emerson, "Design and Synthesis of Synchronization Skeletons Using Branching Time Temporal Logic," *Logic of Programs: Workshop*, LNCS 407, Springer, 1981.

227

[Clarke et al. 2000] E. Clarke, O. Grumberg, and D. Peled, *Model Checking*, The M.I.T. Press, 2000.

[Eijk 1998] C. A. J. van Eijk, "Sequential Equivalence Checking without State Space Traversal," *Proc. Conf. Design, Automation and Test in Europe (DATE 98)*, IEEE CS Press, Los Alamitos, Calif., 1998, pp. 618–623.

[Fallah et al. 1998] F. Fallah, S. Devadas, and K. Keutzer, "OCCOM: Efficient Computation of Observability-Based Code Coverage Metrics for Functional Verification," *Proc. Design Automation Conference*, pp. 152–157, 1998.

[Foster and Coelho 2001] H. Foster and C. Coelho, "Assertions Targeting a Diverse Set of Verification Tools," *Proc. Int'l HDL Conference*, March 2001.

[Foster et al. 2004a] H. Foster, A. Kronik, and D. Lacey, *Assertion-Based Design*, Kluwer Academic Publishers, May 2004.

[Foster et al. 2004b] H. Foster, H. Wong-Toi, C. N. Ip, and D. Perry, "Formal Verification of Block-Level Requirements," *DesignCon*, 2004.

[Gordon 2003] M. J. C. Gordon, "Validating the PSL/Sugar Semantics Using Automated Reasoning," in *Formal Aspects of Computing*, vol. 15, no. 4, December 2003, pp. 406–421, Springer-Verlag, London.

[Hoskote 1999] Y. Hoskote, T. Kam, P-H. Ho, and X. Zhao, "Coverage Estimation for Symbolic Model Checking," *Proc. Design Automation Conference*, 1999.

[Huang et al. 1997] S.-Y. Huang, K.-C. Chen, and K.-T. Cheng, "AQUILA: An Equivalence Verifier for Large Sequential Circuits," *Proc. Asia and South Pacific Design Automation Conf.*, ACM Press, New York, 1997, pp. 455–460.

[Huang 1998] S.-Y. Huang and K.-T. Cheng, *Formal Equivalence Checking and Design Debugging*, Kluwer Academic Publishers, Norwell, Mass., 1998.

[IEEE 1076-1993] IEEE Standard 1076-1993, *VHDL Language Reference Manual*, IEEE, Inc., New York, June 6, 1994.

[IEEE 1364-2001] IEEE Standard 1364-2001, *IEEE Standard Hardware Description Language Based on the Verilog Hardware Description Language*, IEEE, Inc., New York, March 2001.

[Katz et al. 1999] S. Katz, D. Geist, and O. Grumberg, "Have I Written Enough Properties?" in *10th CHARME*, LNCS 1703, pp. 280–297, 1999.

[Kripke 1963] S. Kripke, Semantic Considerations on Model Logic, *Proceedings of a Colloquium: Modal and Many-Valued Logics*, vol. 16 of *Acta Philosophica Fennica*, pp. 83–94, August 1963.

[Kroph 1998] T. Kroph, *Introduction to Formal Hardware Verification*, Springer, 1998.

[Kuehlmann and Krohm 1997] A. Kuehlmann and F. Krohm, "Equivalence Checking Using Cuts and Heaps," *Proc. Design Automation Conf. (DAC 97)*, ACM Press, New York, 1997, pp. 263–268.

[Kunz 1993] W. Kunz, "HANNIBAL: An Efficient Tool for Logic Verification Based on Recursive Learning," *Proc. Intl. Conf. Computer-Aided Design (ICCAD 93)*, IEEE CS Press, Los Alamitos, Calif., 1993, pp. 538–543.

[Loh et al. 2004] L. Loh, H. Wong-Toi, C. N. Ip, H. Foster, and D. Perry, "Overcoming the Verification Hurdle for PCI Express," *DesignCon*, 2004.

[Matsunaga 1996] Y. Matsunaga, "An Efficient Equivalence Checker for Combinational Circuits," *Proc. Design Automation Conf. (DAC 96)*, ACM Press, New York, 1996, pp. 629–634.

[McMillian 1993] K. McMillian, *Symbolic Model Checking: An Approach to the State Explosion Problem*, Kluwer Academic, 1993.

[Moorby et al. 2003] P. Moorby, A. Salz, P. Flake, S. Dudani, and T. Fitzpatrick, "Achieving Determinism in SystemVerilog 3.1 Scheduling Semantics," *Proceedings of DVCon 2003*, 2003. Retrieved March 31, 2003, from the World Wide Web: http://www.eda.org/sv-ec/sv31schedsemantics-dvcon03.pdf.

[PCI-2.2 1998] *PCI Local Bus Specification*, Revision 2.2, PCI Special Interest Group, December 18, 1998.

[Pixley 1992] C. Pixley, "A Theory and Implementation of Sequential Hardware Equivalence," *IEEE Trans. Computer-Aided Design*, vol. 11, no. 12, Dec. 1992, pp. 1469–1494.

[Pixley 1994] C. Pixley et al., "Multi-level Synthesis for Safe Replaceability," *Proc. Int'l Conf. Computer-Aided Design* (ICCAD 94), IEEE CS Press, Los Alamitos, Calif., 1994, pp. 442–449.

[Pnueli 1977] A. Pnueli, "The Temporal Logic of Programs." *18th IEEE Symposium on Foundation of Computer Science*, IEEE Computer Society Press, 1977.

[Pradhan et al. 1996] D. Pradhan, D. Paul, and M. Chatterjee, "VERILAT: Verification Using Logic Augmentation and Transformations," *Proc. Intl. Conf. ComputerAided Design* (ICCAD 96), IEEE CS Press, Los Alamitos, Calif., 1996, pp. 88–95.

[Ritter 2000] G. Ritter, "Sequential Equivalence Checking by Symbolic Simulation," *Proc. 3d Int'l. Conf. Formal Methods in Computer-Aided Design* (FMCAD 2000), Lecture Notes in Computer Science, vol. 1954, Springer-Verlag, New York, 2000, pp. 423–442.

[Roth 1977] J. P. Roth, "Hardware Verification," *IEEE Trans. Computer-Aided Design*, vol. 26, no. 12, Dec. 1977, pp. 1292–1294.

[Zhang and Malik 2002] L. Zhang and S. Malik, "The Quest for Efficient Boolean Satisfiability Solvers," in Ed Brinksma and Kim Guldstrand Larsen, eds., *Proc. Computer Aided Verification*, vol. 2404 of LNCS, pp. 17–36, Copenhagen, Denmark, 2002.

INDEX

ABOUT THE AUTHORS

Douglas L. Perry is the Director of Marketing for Virtutech, Inc. He is the author of four editions of McGraw-Hill's *VHDL*. He lives in San Ramon, California.

Harry D. Foster serves as Chairman of the Accellera Formal Verification Technical Committee, which is currently defining the PSL (Property Specification Language) standard. He is co-author of the new Kluwer Academic Publishers book *Assertion-Based Design*. The Chief Methodologist at Jasper Design, Mr. Foster formerly was Verplex Systems' Chief Architect. He lives in Richardson, Texas.